P9-BZZ-846

STP 1308

Testing and Acceptance Criteria for Geosynthetic Clay Liners

Larry W. Well, editor

ASTM Publication Code Number (PCN):
04-013080-38

ASTM
100 Barr Harbor Drive
West Conshohocken, PA 19428-2959
Printed in the U.S.A.

ISBN: 0-8031-2471-6
PCN: 04-013080-38

Peer Review Policy

Each paper published in this volume was evaluated by two peer reviewers and at least one of the editors. The authors addressed all of the reviewers' comments to the satisfaction of both the technical editor(s) and the ASTM Committee on Publications.

To make technical information available as quickly as possible, the peer-reviewed papers in this publication were prepared "camera-ready" as submitted by the authors.

The quality of the papers in this publication reflects not only the obvious efforts of the authors and the technical editor(s), but also the work of these peer reviewers. The ASTM Committee on Publications acknowledges with appreciation their dedication and contribution of time and effort on behalf of ASTM.

Printed in Philadelphia, PA
January 1997

Foreword

The Symposium on Testing and Acceptance Criteria for Geosynthetic Clay Liners was held 29 January 1996 in Atlanta, Georgia. ASTM Committee D35 on Geosynthetics through its Subcommittee D35.04 on Geosynthetic Clay Liners sponsored the symposium. Larry W. Well, CH2M Hill, Inc., presided as symposium chairman and Kent von Maubeuge, Naue Fasertechnik Gmbh and Company, presided as symposium cochairman. Larry W. Well is editor of this publication.

Contents

Overview

A symposium sponsored ASTM Subcommittee D35.04 on Geosynthetic Clay Liners was held on 29 Jan. 1996 at the Hyatt Regency in Atlanta, Georgia. The intention of the program was to bring together the current knowledge and understanding about the use of this relatively new product currently being used in containment systems. The symposium had attendees from around the world and presenters from Canada, France, Germany, and the United States. The international flavor of the symposium and technical papers suggests the world has embraced this new product and its representatives are eager to share their experience and knowledge. This cooperation is a very positive signal that the development of single and composite containment systems using GCLs will continue well into the next millennium. There is no doubt that additional forums about GCLs will be held in the future to add to the knowledge base we currently have.

Background

Geosynthetic clay liners (GCLs) are developing as an important and accepted component in hydraulic barriers for containment systems. Used in conjunction with geomembranes they are becoming commonplace, working as partners to make reliable composite lining systems for landfill lining and capping systems. GCLs alone are also effective barrier linings for liquid impoundment. The fact that they are manufactured products made under comprehensive quality control and fabrication standards provides a performance reliability that compacted clay soil layers do not often achieve.

The mid-1980s saw the introduction of GCLs into the containment industry. Since that time there have been millions of square meters of GCLs installed. The designs for these innovative installations involve careful review of the basic soil mechanics and engineering principles to accomplish design goals. Design goals are related to hydraulic conductivity, chemical resistance, slope stability, uniformity of product, environmental influences, and installation cost. The careful application of the scientific and engineering principles in GCL applications has fortunately prevented many significant failures. A basic tool in the workshop of engineering is past experience to learn what works well and what is not successful. With over a decade of practice using GCLs the time was ripe to hold this symposium to examine where we have been, what we have to work with now, and chart our course for the future.

Organization

The Special Testing Publication (STP) No. 1308 "Testing and Acceptance Criteria for Geosynthetic Clay Liners" presents state-of-the-art review. A summary is provided of the types of products, special test procedures developed over the last several years, and results of studies of hydraulic, chemical compatibility and stability issues. Experiments with various chemicals, and case histories are also presented to inform the reader of the state of the practice. The STP is arranged to present a progression from special testing procedures, shear strength and creep testing results, hydraulic conductivity testing, chemical compatibility issues, and, finally, to specification suggestions for GCL applications. We hope the information contained herein provides insight, guidance, and confidence in proper applications of GCLs in containment systems for liquids, solid, and hazardous waste containment systems. We intended for the STP to be used as a reference in future work for designers, regulatory agencies, and professors as the topic of geosynthetic containment systems becomes more common place in college curricula. Currently there does not appear to be a host compendium for information about GCLs, and we intend this STP to be a beginning for a collection of pertinent technical information.

Contents

Keynote Address

The Keynote Address **Perspectives of Geosynthetic Clay Liners** presents a neat and tidy overview that addresses a number of topics and issues pertaining to geosynthetic clay liners. The relevant physical, hydraulic, mechanical, and endurance properties, and their associated test methods are reviewed for the perspective of manufacturing quality control tests versus design oriented or performance tests. Current activity and recommendations are suggested for these relatively new and unique liquid barrier materials.

Special Testing Procedures

The paper that describes the **Effect of Moisture Content on Free Swell of the Clay Component of Geosynthetic Clay Liners** discusses the effect that initial moisture content has in causing the free swell to decrease linearly, in some GCLs, and nonlinearly in with other GCL products. A procedure for calculating the free swell from the initial moisture content is presented.

The **Effect of Swell Pressure on GCL Cover Stability** describes the importance of bentonite swell pressure in the stability of a soil cover system that incorporates a GCL. The typical one-dimensional swell test indicates greater swell pressure than typical overburden loading provides. Hence, slope stability may be compromised. The paper also presents an innovative suggestion for development of GCL products to increase cover stability.

A Comparison of Sample Preparation Methodology in the Evaluation of Geosynthetic Clay Liner (GCL) Hydraulic Conductivity presents the important procedures for taking testing specimens used for hydraulic conductivity testing from GCL samples. Different conditioning and trimming procedures are examined and rated for effectiveness.

Shear Strength and Long-Term Creep Testing

The **Effect of Normal Stress During Hydration and Shear on the Shear Strength of GCL/Textured Geomembrane Interfaces** used two pre-shear inundation methods designed to simulate field conditions in a laboratory testing program. The purpose is to evaluate the interface shear strength. The results of two different GCL materials and the two pre-shear conditions are compared.

A design case history examined one GCL product in **Strength and Conformance Testing of a GCL Used in a Solid Waste Landfill Lining System.** As part of a composite landfill lining system in Seismic Zone 4 area on a relatively steep slope required careful evaluation of both internal shear strength of the GCL and the interface friction between the GCL and textured HDPE liner. Stability analyses using stress-dependent interface and internal shear strengths for the GCL are discussed. Quality assurance and conformance testing for the project on the GCL are also discussed.

The **Short-Term and Creep Shear Characteristics of a Needlepunched Thermally Locked Geosynthetic Clay Liner** are explored by a series of constant-rate direct shear tests following procedures of ASTM D 5321. The test results demonstrate that the needlepunched thermally locked reinforcing fibers provide substantial short-term shear strength to a GCL. The long-term shear strength can be affected due to potential creep within the reinforcing fibers and this was examined in a newly developed constant-load (creep) shear testing device.

Another study on the **Long-Term Shear Strength Behavior of a Needlepunched Geosynthetic Clay Liner** describes two large scale constant-load (creep) shear testing devices that were developed to evaluate shearing behavior and interfaces between GCLs and other geosynthetics or soils. One device simulates loading conditions that typically occur in a cover system. The other device simulates loading conditions that may occur on a GCL deployed in a landfill lining system. Further testing is planned to more accurately define the time-dependent internal and interface shear behavior of the GCL.

In another paper guidelines are presented for **Shear Strength Testing for Geosynthetic Clay Liner** for measuring the internal and interface shear strengths in a direct shear test. Currently, there is a significant variability between laboratories in shear testing procedures in relation to specimen hydration and shear rate. The intent of the study is to present methods to provide measured strengths that are representative of typical field conditions.

Hydraulic Conductivity Testing Compatibility Issues

A design case history is used to describe the rationale used to evaluate the use of **Geosynthetic Clay Liners in Alkaline Environments.** Designers are faced with equivalency calculations for comparison of GCL performance with compacted clay liners. One important property is the ability of the hydraulic conductivity of the GCL to withstand degradation due to permeation of alkaline leachates rich in materials such as calcium. The paper discusses the concept of ion exchange and the resulting decrease in swell potential, and adsorption capacity and the increase in hydraulic conductivity. Conclusions suggest appropriate performance testing with the leachate in question must be performed during the design phase and confirmed by construction quality assurance testing.

Laboratory Simulation of Geosynthetic Clay Liner Application in Contaminated Liquids Evacuation evaluated the ability of a GCL in preventing leakage of mineral oil into ground water. A full-scale model was designed and constructed for the testing program that showed only one-percent of the precipitated water and leaked mineral oil was collected underneath the GCL. Further research is recommended on techniques of seaming GCLs; the minimum acceptance rate of hydration of GCLs for different liquids; the influence of the water content of adjacent soils in GCL hydration; and long-term hydraulic compatibility of GCLs with different liquids and leachates.

Experiments were conducted to investigate **First-Exposure Performance of the Bentonite Component of a GCL in a Low-pH Calcium-Enriched Environment** by testing the compatibility of the sodium-bentonite component subjected to acidic ground water. A second test was performed to learn the combined effects of acidic ground water enriched with calcium. The relationship between the ionic exchange and changes in hydraulic conductivity and electrical conductance are reported and discussed.

Hydraulic Conductivity Testing Issues and Methods

The **Influence of Initial Hydration Conditions on GCL Leachate Permeability** describes the climatic situations occurring between installation of the GCL and installation of waste in a landfill. The first incident of leachate exposure can correspond to various degrees of hydration of the GCL. The paper aims at analyzing the partially and totally hydrated GCLs behavior after a long-term exposure to leachate. Tests were performed on two needle-punched GCLs in three conditions of hydration that show an important variation of the permeability related the degree of initial hydration.

In the continuing evolution of testing procedures **Measurement of Hydraulic Properties of Geosynthetic Clay Liners Using a Flow Box** is described and evaluated. The GCL flow box offers several advantages over large tanks or flexible wall pemeameters. The conclusions of the testing programs on large-scale intact specimens, overlapped seams, and GCLs under environmental stresses such as freeze-thaw is that the tests may be performed more conveniently and reliably for verifying hydraulic conductivity of GCL panels.

The most important variables in hydraulic conductivity testing are addressed in **Laboratory Hydraulic Conductivity Testing of GCLs in Flexible-Wall Permeameters**. The paper describes the variables and the round-robin testing program conducted by 18 laboratories to independently measure the hydraulic conductivity of a GCL permeated with water. All test specimens came from the same sample. The coefficient of variability for the round-robin testing program, for experienced laboratories, and for a manufacturers 7-month quality control testing period are presented. The results of the round-water testing are encouraging, considering there was less variability than might be expected.

Specifications for GCL Applications

Geosynthetic clay liners are being used on increasingly steep and high landfill slopes, which requires careful determination of the appropriate shear strength to be used in design. **A Design Perspective on Shear Strength Testing of Geosynthetic Clay Liners** reviews shear strength testing of GCLs and the subsequent use of the strength data, from a design viewpoint. There are references to landfills in Hong Kong's mountainous terrain that has many steep-sided valleys or canyons with 25 to 40° natural slopes rising from near sea-level to a few hundred meters.

A discussion of **Manufacturing Quality Control and Specification Criteria for Geosynthetic Clay Liners** is aimed at state-of-the-art quality control and quality assurance testing to help maximize GCL performance. Newly developed ASTM standards for GCL testing and practice are described as they closely relate to quality control and development of useful specifications for construction projects.

A Study of the CBR Bearing Capacity Test for Hydrated Geosynthetic Clay Liners describes a modified version of the California Bearing Ratio (CBR) penetration test used to investigate hydrated GCL bearing capacity. Comparisons of various styles of GCL products are presented as influenced surcharge pressure, displacement rate, mold diameter, and test termination criteria based on lateral bentonite migration (squeezing). Recommendations for equipment and methodologies for the test are included in the paper.

Conclusion

Some very good work has been done in the last decade in developing, using, testing, and understanding geosynthetic clay liner materials. The geosynthetics industry is one of innovation and response to critical needs in many areas of remediation, infrastructure repair and construction, and the containment of liquids, solids, and hazardous wastes. The construction market place is open to innovation. One of the goals of the symposium and this STP is to present a kernel of knowledge and insight into the understanding of a unique and special product. Hopefully, this kernel will germinate and continue to grow and expand the knowledge and experience base we will rely on in the coming years. There is much important work yet to be accomplished. The purpose and topics of the symposium and this STP will serve us well if they stimulate further research and development of the product, provide a basis for effective applications, and expand the geosynthetics industry. We trust the results will enhance our environmental well-being.

Larry W. Well

CH2M Hill, Inc.,
Portland, Oregon;
editor.

Keynote Address

Robert M. Koerner[1]

PERSPECTIVES ON GEOSYNTHETIC CLAY LINERS

REFERENCE: Koerner, R. M., ''**Perspectives on Geosynthetic Clay Liners,**'' *Testing and Acceptance Criteria for Geosynthetic Clay Liners, ASTM STP 1308,* Larry W. Well, Ed., American Society for Testing and Materials, 1997.

ABSTRACT: This overview paper addresses a number of topics and issues pertaining to geosynthetic clay liners, or as they are usually referenced, GCLs. GCLs are factory manufactured products consisting of a layer of bentonite contained between two geotextiles by using adhesives, stitch bonding or needle punching, or by adhesively bonding the bentonite to a geomembrane. A number of commercially available products are available on a worldwide basis. The various products are reviewed along with some of the basic elements of their manufacturing process. The primary areas of acceptance and utilization are also presented.

Relevant physical, hydraulic, mechanical and endurance properties and their associated test methods are presented from the perspective of manufacturing quality control (MQC) tests versus design oriented (or performance) tests. Research and development testing is also addressed as contrasted to MQC and design testing. Lastly, current activity and recommendations are suggested for these relatively new and unique barrier materials.

It is hoped that the paper sets the proper tone for subsequent papers in this symposium which will add to the growing body of information and knowledge. Indeed, GCL's (acting individually or as composite materials) are changing the very essence of liquid barriers and their associated containment systems.

KEYWORDS: Geosynthetic clay liners, GCLs, bentonite, geotextiles, geomembranes, adhesives, needle punching, stitch bonding, heat burnishing, GCL products, GCL manufacturing quality control testing, MQC, design tests, performance tests, R & D tests

---

Geosynthetic clay liners, or GCLs, consist of factory manufactured rolls of bentonite between geotextiles, or on a geomembrane. The bentonite is the low hydraulic conductivity (or permeability) natural soil component, while the geosynthetics act as support and reinforcement materials. The geosynthetics also provide

---

[1]H. L. Bowman Professor of Civil Engineering and Director of the Geosynthetic Research Institute, Drexel University, Philadelphia, PA 19104.

the opportunity to stitch bond, needle punch or adhesively bond the bentonite into a unit suitable for handling, transportation and placement as a composite barrier material. Even further, the geosynthetics provide the mechanism by which shear strength can be provided within the mid-plane of the as-installed products when used on side slopes and in various other configurations.

BACKGROUND OF GCLs

This section briefly describes the mineral bentonite and how it is used and configured in geosynthetic clay liners (GCLs). It also presents a brief history of the development of the various GCL products and the currently available types and styles.

Bentonite

Clay minerals are the smallest of geologically occurring materials and are characterized by their thin layered structure. They are negatively charged on their surfaces which allows for the attraction of a polar fluid like water. The bound water layers (together with partially hydrated cations) occupy free void volume between the soil particles thereby reducing the hydraulic conductivity of the composite soil/water system. Among the most active of the clay minerals is montmorillonite. Montmorillonite has highly negative charged surfaces as evidenced by its high cation exchange capacity, see Table 1. This leads to a unique set of properties including an extremely low hydraulic conductivity, or permeability.

TABLE 1--Base exchange capacity of bentonites after Madsen and Nüesch [1].

	Na-bentonite MX-80*	Ca-bentonite Montigel**
Montmorillonite content [% by weight]	75	66
Specific surface area [m^2/g]	560	490
Exchange capacity [meq/100g]	76	62

*from Wyoming, USA
**from Bavaria, FRG

Bentonite is the most common member of the montmorillonite group. It is mined from naturally occurring formations. As such, it often contains other minerals such as quartz, feldspar, mica and other clay minerals. High quality bentonite contains 65 to 95% montmorillonite by weight, Egloffstein [2]. For industrial applications like GCLs, bentonite can be classified into three types:

- natural sodium bentonite
- natural calcium bentonite
- soda activated bentonite

Natural sodium bentonites are found in Wyoming and the Dakotas in North America, while natural calcium bentonites are found in many European countries and elsewhere. Since calcium bentonites have a higher hydraulic conductivity than sodium bentonites, the soda activated types (calcium ions replaced by sodium ions) are usually used for barrier, or sealing, applications instead of natural calcium bentonites. Thus, GCLs consist of either natural sodium bentonites or soda activated bentonites as the low hydraulic conductivity component of the material. For a more complete review of bentonites used in the manufacture of GCLs see Trauger [3] and the associated references.

The Earliest GCLs

Stemming from the long success of a rigid panel-type foundation waterproofing product called Volclay® panels, the transition into a flexible product capable of being rolled was made in 1982. Significant modifications were the type of adhesives, the processing (particularly the oven heating) and the use of geotextiles in place of cardboard support layers. The resulting product, Claymax®, was used as early as 1986 in double lined landfill liner systems in the USA. In particular, when the GCL was placed beneath the primary geomembrane of a liner system subsequent leakage into the underlying drainage system was virtually eliminated as witnessed by sampling in the leak detection sump.

In approximately the same time period, the product Bentofix® was developed in Germany. It uses bentonite powder placed between two needle punched geotextiles and then needles the entire composite together. This needle punched GCL was successfully used in the 1980's in a variety of applications such as canal liners, pollution prevention in airfields and highways, and other barrier applications.

Elements of GCL Manufacturing

Upon mining the bentonite from the geographic area where it is located it is oven dried to remove field moisture. A sorting and/or sieving of the impurities is then undertaken. Final sizing takes place and the bentonite is tested for various quality control parameters. It is then transported to the GCL manufacturing facility where additional quality control tests may be performed. If acceptable, the bentonite is stored in silos until ready for use. It can be in powder or granular form.

Figure 1 shows schematic diagrams of the two basic manufacturing processes for the GCLs just described. The upper sketch illustrates the multi-stage placement of bentonite and adhesive followed by oven heating for drying of the product. There is a nominal moisture content remaining due to residual adhesive and the tendency of the bentonite to absorb humidity from the atmosphere. The amount is product dependent and varies from 15 to 25%, or more. The lower sketch of Figure 1 illustrates the manufacture of a needle punched GCL where the bentonite layer, sometimes consisting of different particle sizes, is placed on a carrier geotextile with a covering geotextile placed above. It then

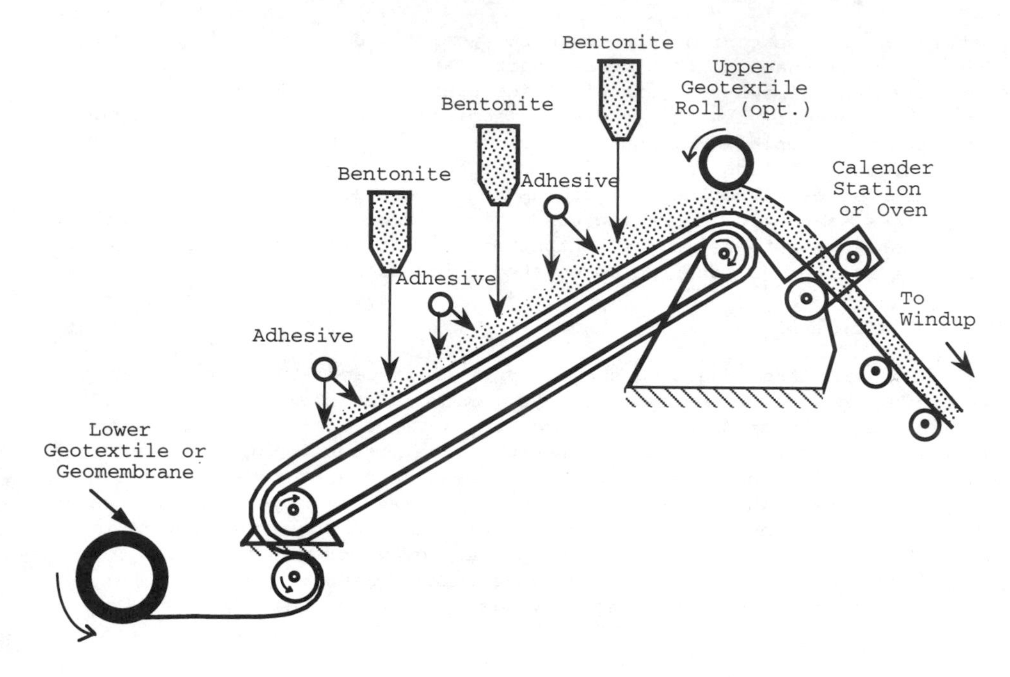

(a) Adhesive mixed with clay

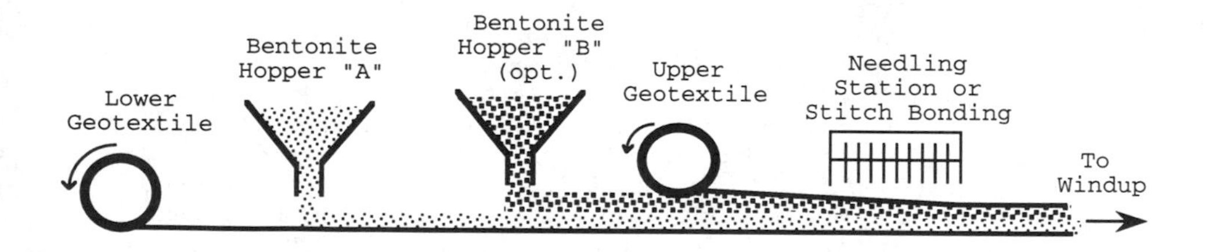

(b) Needle punched or stitch bonded through clay

FIG. 1--Schematic diagram of manufacture of different types of Geosynthetic Clay Liners (GCLs), after Koerner [4].

passes through a needling station. The geotextile on the surface facing the needles must be of the needle punched type since the needles during GCL manufacture intercept and drag fibers through the bentonite and underlying geotextile. It should be noted that needles for GCL manufacture are considerably longer and stronger then those used in the manufacture of light weight needle punched geotextiles. The final GCL product has a residual moisture content which depends on the local humidity, fineness of bentonite, etc. It is usually in the range of 10 to 15%. The manufactured GCLs are then wound on a rigid core to the desired length and weight, and wrapped in a polyethylene film or a tight fitting polymer bag for storage and transportation.

Current GCL Types and Styles

Since the introduction of the initial adhesive and needled GCL products just described, many other concepts and modifications of GCL manufacturing have been developed. Some of these are the following:

- The practice of stitch bonding GCL products to add mid-plane shear strength is available in both the Claymax® and NaBento® products. The stitching process is described by Fuller [5].
- The use of a geomembrane as a carrier material for a layer of adhesively bonded bentonite results in a product called Gundseal®. It is a composite geomembrane/ bentonite GCL. Its manufacture is described by Erickson and Anderson [6].
- The locking of the fibers which penetrate through the thickness of needle punched GCLs is available on Bentofix® and Bentomat® products by either thermal means (heat burnishing) or by chemical bonding, see von Maubeuge and Heerten [7].
- The bentonite itself is sometimes in a powder state, but can also be in a granular form and can even be placed as granules with powder as the infill.
- An open-structure nonwoven interlayer can be used to avoid shifting of the bentonite powder during transport and installation and to aid in stabilization.
- A wide variety of geotextiles are used as carrier materials (substrate and/or supersubstrate) such as, needle punched nonwovens, woven slit films, woven composites, and spun laced fabrics.
- A wide variety of geomembranes can be used as substrate materials, such as HDPE, VLDPE or LLDPE, either smooth or textured.
- An even greater variety of geomembranes can be field placed as superstrate materials so as to maintain low moisture content in the encapsulated bentonite layer.
- New GCLs such as a prehydrated product (albeit at a low moisture content), a combined needle punched/heat bonded product and a geogrid/geonet supported composite are being developed in the United Kingdom, Austria and the USA, respectively.

As a result of this innovative manufacturing activity, the designer has available a number of different types of GCLs. They are shown in the cross sections of Figure 2. Furthermore, within each of the cross

(a) Adhesive Bound Clay to Upper and Lower Geotextiles

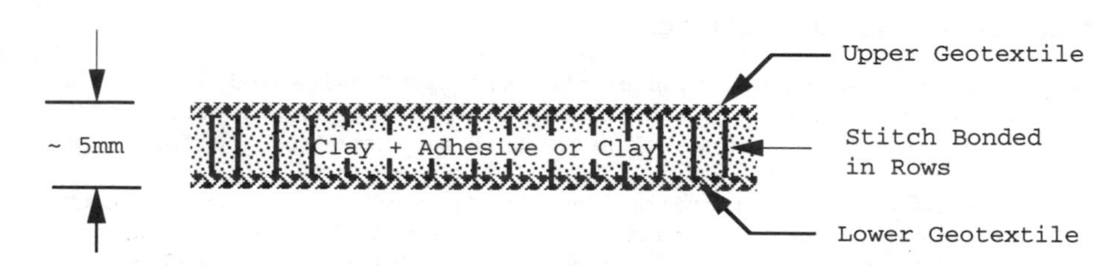

(b) Stitch Bonded Clay Between Upper and Lower Geotextiles (sometimes with an additional interlayer)

(c) Needle Punched Clay Through Upper and Lower Geotextiles

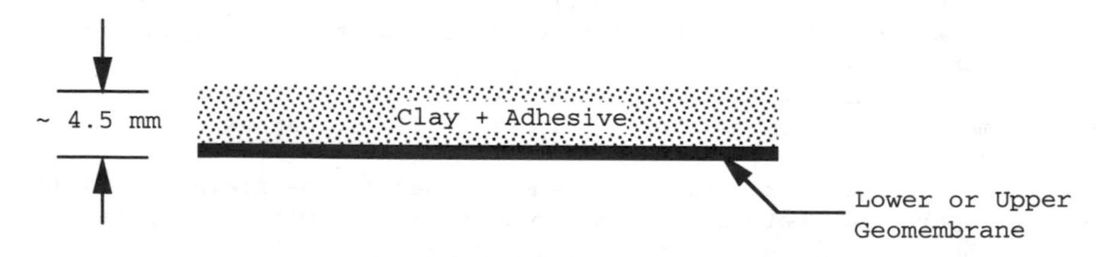

(d) Adhesive Bound Clay to a Geomembrane

FIG. 2--Cross section sketches of currently available geosynthetic clay liners (GCLs).

sections shown most manufacturers have different styles available, see Table 2. It should be noted that the situation is one of rapid change and evolution, thus the manufacturers should be consulted as to possible product changes and modifications.

ACCEPTANCE AND UTILIZATION

The acceptance of GCLs has been very strong among private owners of landfills, surface impoundments and waste piles. The general application is for facilities in need of barrier materials to either prevent liquids from leaking out of, or from entering into, a given containment system. From a regulatory perspective, there have been three U.S. Environmental Protection Agency (EPA) sponsored workshops on GCLs which have presented the majority of available research and development information in the USA, see Daniel, et al. [8, 9, 10]. From application and testing perspectives there have been two conferences; one in Germany [Koerner, Gartung and Zanzinger [11]] and this American Society for Testing and Materials (ASTM) Conference [Well and von Maubeuge[12]].

From this growing body of information (conferences, symposia, workshops, journals, reports and magazines) the applications that GCLs serve can be classified into environmental, transportation and geotechnical categories.

Environmental Applications

The principal environmental applications for GCLs are focused on solid waste containment, i.e., landfills. There are three locations where GCLs have been used. These are in the cover, the lower component of the primary liner and the lower component of the secondary liner. Figure 3 illustrates GCLs in these three locations. Note that they are illustrated as composite geomembrane/GCL systems but this is not always necessary. In some situations GCLs can be used as a single component liner, i.e., by themselves. Even further the secondary liner in Figure 3 is illustrated as being a three-component composite, i.e., geomembrane, GCL and natural soil, albeit the natural soil is of higher permeability than the typically regulated value of 1×10^{-7} cm/sec for compacted clay liners. These particular application areas are arguably the largest current use of GCLs, at least in North America.

A related environmental application area is the prevention of seepage from surface impoundments containing contaminated liquids. In this case, GCLs are generally used as a composite liner with an overlying geomembrane. Secondary containment for underground storage tanks have utilized GCLs in a number of cases. The innovative use of GCLs as groundwater protection from de-icing chemicals at Munich's airport is well documented, Scheu, et al. [13].

Transportation Applications

The primary area of GCL use in transportation facilities is the

TABLE 2--Currently available GCLs in North America. [Note, however, that the situation changes on a regular basis and the respective manufacturers should be consulted in this regard].

Manufacturer/ Product Trademark	GCL Style	Bentonite Type*	Bentonite Form	Upper Geotextile	Lower Geotextile	Bonding Method
Albarrie Naue Ltd./ Bentofix® Thermal Lock	NS	sodium	granules	woven	nonwoven	needle punched & heat burnished
	WP	sodium	granules	woven	nonwoven	needle punched & heat burnished
	NW	sodium	granules	nonwoven	nonwoven	needle punched & heat burnished
CETCO/Claymax Div. Claymax®	200R	sodium	granules	woven	woven	adhesive
	500 SP	sodium	granules	woven	woven	adhesive & stitch bonded
Colloid Env. Tech. Co. (CETCO)/Bentomat®	—	sodium	granules	woven	nonwoven	needle punched
Gundle Lining System/ Gundseal®	—	sodium	granules	none	polyethylene geomembrane	adhesive
Huesker Inc./ NaBento®	—	sodium or soda activated	powder	nonwoven	nonwoven	stitch bonded & fabric interlayer

*most GCLs have 5.0 kg/m^2 (1.0 lb/ft^2) of bentonite as the targeted mass per unit area

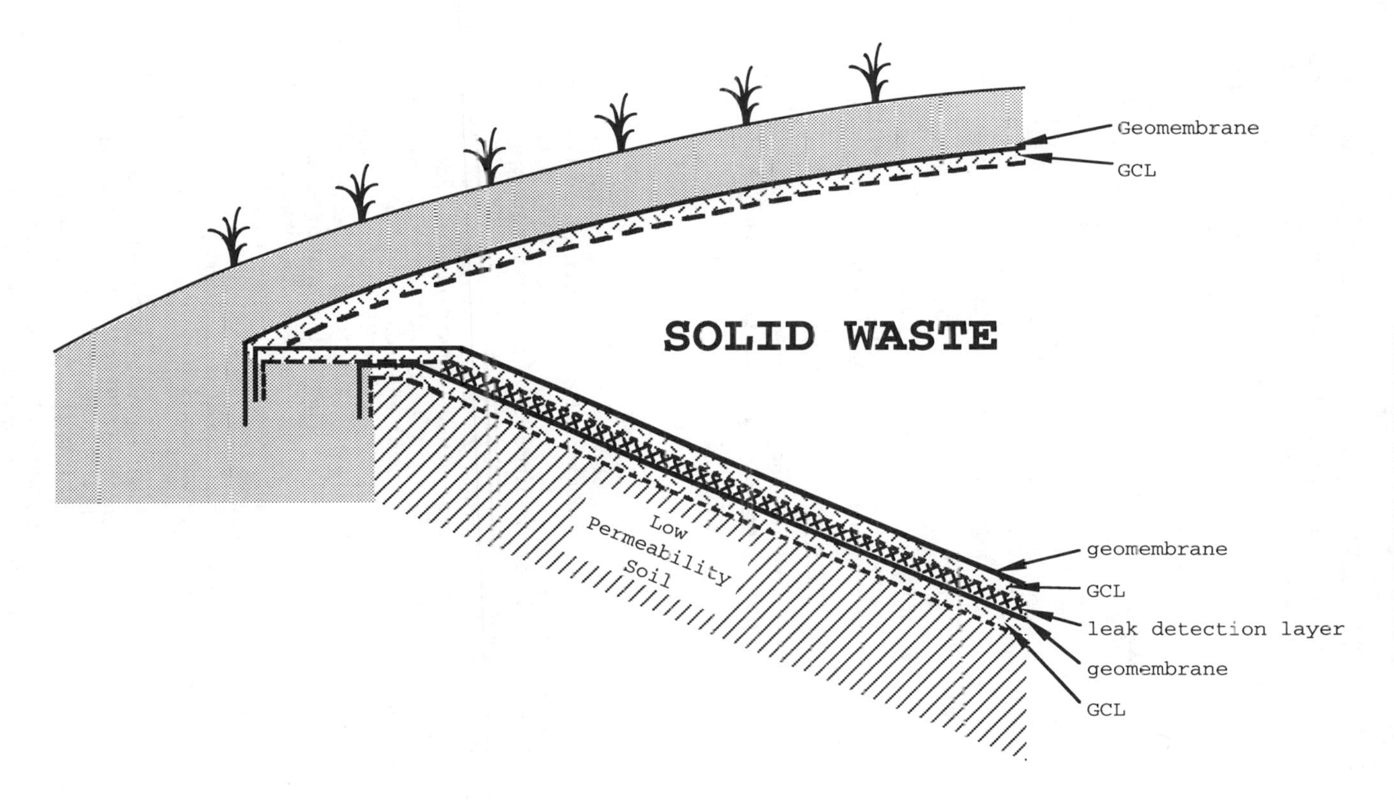

FIG. 3--Primary use of GCLs in the waste containment systems of landfills.

prevention of pollution of sensitive subsurface strata from truck spillage and accidents, see Rathmayer [14]. Additionally, GCLs have been used where salt is used on roadways and the local groundwater is to be protected against pollution, see Schmidt [15]. While not known to have been used as such, railroad refueling depots could utilize GCLs for environmental protection and reclamation of spillage in a similar manner as geomembranes.

Geotechnical Applications

The control of seepage in earth, earth/rock and masonry dams is of major importance. Using barriers such as GCLs not only achieves this goal, they also decrease or eliminate excess pore water pressures thereby increasing stability. Perhaps the most innovative use of GCLs in this regard is the study of Rau and Dresher [16] on the upper basin of a pumped storage power station in Germany.

Canal liners have utilized GCLs as reported by Heerten and List [17] and can possibly be used in an exposed condition, i.e., not soil backfilled. A laboratory feasibility study by Koerner, G. R., et al. [18] is available.

GCLs have also been used in vertical cutoff walls, Heerten, et al. [19]. The specific gravity of GCLs is such that submergence in a slurry supported trench was readily achievable.

TEST METHODS AND PROPERTIES

Considering the historic use of a material like bentonite, coupled with manufactured geosynthetic materials such as geotextiles and geomembranes, it should come as no surprise that there are a large number of test methods that can be used to characterize the properties of the bentonite, the geosynthetics and the composite GCL. This tendency toward testing for all possible properties, however, should be controlled to the point that the required properties are indeed meaningful for the intended purpose and/or application. To test GCLs for the sake of data, per se, is excessively costly and possibly counterproductive. In this section an attempt is made to present the most relevant properties (and their associated test methods, when available) from the perspectives of manufacturing quality control (MQC) and design. Conformance testing, e.g., to verify that the proper material has been brought to the job site, will typically be a subset of MQC testing albeit at a reduced testing frequency. This is obviously a site specific decision. These same comments also apply to manufacturing quality assurance (MQA) testing.

Regarding manufacturing quality control (MQC), there are an enormous number of tests that might be utilized. Those that appear to be somewhat relevant appear in Table 3. The list is seemingly excessive but it should be noted that the geotextiles or geomembranes that are used in GCL manufacturing have all of the listed tests already performed on a routine basis in their own manufacturing. Thus the information is readily available. The bentonite tests are for the bentonite powder or granules and the composite GCL is for the final product. This type of categorization is also found in Heerten, et al. [19] and in ASTM D5889 which is focused on MQC testing of GCLs.

TABLE 3-- Manufacturing Quality Control (MQC) guide modified and expanded from Heerten, et al. [20]. [Note that ASTM has a related guide under the designation of D5889].

Property	Test Method	Limiting Value	Frequency of Testing
Bentonite:			
Moisture content (max.)	ASTM D4643	10%	Based on specific shipment unit or 50 tonnes
Swell index[1] (min.)	ASTM D5890	25 ml	
Moisture adsorp. (min.)	Enslin-Neff[2,4]	600%	
Geotextile:[3]			
Mass/unit area[1]	ASTM D5261	product specific	Typically 20,000 m^2 (200,000 ft^2)
Thickness	ASTM D5199	" "	
Grab Tensile Strength[1]	ASTM D4632	" "	
Trap. Tear Strength	ASTM D4533	" "	
Burst Strength	ASTM D3786	" "	
Puncture	ASTM D4833	" "	
Geomembrane:[3]			
Thickness (smooth)[1]	ASTM D5199	product specific	Typically 20,000 m^2 (200,000 ft^2)
Thickness (textured)[1]	GRI GM8	" "	
Tensile Strength[1]	ASTM D638	" "	
Tear Resistance	ASTM D1004C	" "	
Puncture	FTM 101C 2065	" "	
GCL:			
Mass/unit area[1] (min.)	ASTM D5261	5.0 kg/m^2 (1.0 lb/ft^2)	Frequency varies greatly (being currently balloted in ASTM D35.04)
Thickness	ASTM D5199	product specific	
Grab Tensile Strength	ASTM D4632	" "	
Puncture Resistance	ASTM D4833	" "	
Peel Strength[5]	ASTM D4632	" "	
Moisture Content[1] (max.)	ASTM D4643	25%	
Flux (max.)[1]	ASTM D5887[2]	product specific	

[1]Currently under consideration for a ASTM MQC Guide Standard
[2]Using distilled, deionized water
[3]Properties are evaluated on the material before manufacturing into the GCL product. All of the tests listed are routinely performed on the respective manufactured geotextiles or geomembranes.
[4]See Koerner [4] for test method description
[5]Applicable only for needle punched GCLs

Design (Performance) Tests

On the issue of a recommended guide for design tests little can be offered insofar as firm recommendations, since site specific conditions dictate each situation. The design engineer is critical in this decision process. Some of the more widely used tests are presented in Table 4 with commentary as to where the resulting properties may be of importance. In all cases, the specific limiting values and the frequency of testing is also site specific and essentially at the discretion of the design engineer.

TABLE 4--Selected design (performance) tests for GCLs.

Property	Test Method	General Comments
Flux	ASTM D5887	• generally always important • should use site-specific permeant • should use site-specific stress and pressure conditions • should have thickness measured so as to calculate hydraulic conductivity (permeability)
Direct Shear	ASTM D5321	• necessary for most side slope designs • generally upper interface is of main concern • sometimes mid-plane is required • sometimes lower interface is required
Creep Shear	ASTM D5321-mod.	• difficult and costly test • sometimes necessary with low quasi-static FS values • generally upper interface is of concern • sometimes mid-plane is required • rarely lower interface is required
Wide Width Tensile Strength and Elongation	ASTM D4595	• only necessary when tensile stresses are to be resisted • possibly when shear stresses are to be resisted
Wide Width Tensile Strength for Overlaps	ASTM D4595-mod.	• only necessary when shear stresses are to be resisted which include overlaps
Multi-Axial Tension	ASTM D5617-mod.	• for anticipated yielding subgrade situations
Soil Compatibility, or Indentation	ASTM D5818-mod.	• for subgrades with soil particles ≥ 12 mm (0.5 in.) • for cover soils with particle sizes ≥ 12 mm (0.5 in.)
Chemical Resistance	ASTM (in progress)	• for aggressive liquid permeants

Research and Development Tests

Note that Tables 3 and 4 could have contained many additional tests that are referenced in the technical literature, e.g.,

- overlap seam permeability
- transmissivity of upper geotextile (intimate contact issue)
- bearing capacity of the hydrated product
- shrink-swell behavior
- freeze-thaw behavior
- soil subgrade suction tests
- overlap seam shear strength
- water breakout time
- solute breakout time
- chemical adsorption capacity

These tests are excellent research and development tests and give significant insight to the general behavior of GCLs. However, they only occasionally reflect a uniqueness to a particular site specific situation. Thus it is felt that they should not be cited in specifications on a routine basis, but should be very selectively used.

CURRENT ACTIVITY AND FUTURE PROJECTIONS

The GCL community (consisting of manufacturers, designers, testing laboratories, owner/operators, regulators and researchers) is actively investigating a number of facets of these unique engineering barrier materials.

Test Method Development

Under the chairmanship of Mr. Larry W. Well, ASTM has a specific subcommittee, D35.04, focused on GCL test method development. This includes the development of test methods, specifications, guides and standards of practice. There are separate task groups working on physical/ mechanical, hydraulic, endurance and logistics draft standards at the present time.

Other standardization groups such as the International Standards Organization, the European Standards Board, and many national standards setting groups are pursuing similar goals. Hopefully, the resulting standards of these various groups will be interchangeable when fully developed.

Designing with GCLs

GCLs represent a hybrid construction material consisting of soils and geosynthetics. As such, geotechnical engineers are generally more familiar and comfortable with the idiosyncrasies of their design than with other strictly polymeric geosynthetic materials. For example, both hydraulic conductivity and direct shear testing (the two fundamental design issues) are reasonable extensions of familiar geotechnical test methods. For example, ASTM D5084 is the flexible wall permeameter test for determining the hydraulic conductivity of compacted clay liners and ASTM D3080 is the traditional direct shear test used in all soil testing laboratories. Clearly, designing with GCLs in the context of these test methods is within the geotechnical state-of-the-practice. The extension that is required of the design engineer is that needling and/or stitch bonding provides the long term shear strength necessary to augment the relatively low shear strength of hydrated bentonite. Various papers in this conference address these specific issues.

Field Performance of GCLs

To date, the performance assessment of GCLs has been the result of field use and feedback. For example, Weiss, et al. [21] have evaluated the field permeability of GCLS with overlaps under simulated waste subsidence with no significant differences from locations without overlaps. Also, GCLs placed under geomembranes in the primary liner systems of double lined landfills (as shown in Figure 3) have had an excellent record of minimizing leakage in the underlying drainage systems, GeoSyntec [22]. There are also plans to use lysimeters beneath landfill caps containing GCLs, but data from this application are not yet available.

On the shear strength issue, several test plots have been constructed by private owner/operators with generally good success. However, the largest field effort is the fourteen full scale test plots in Cincinnati, Ohio constructed under the auspices of the U.S. Environmental Protection Agency (EPA). These full scale test plots are at both 3 to 1 and 2 to 1 slopes, i.e., 18 and 26 deg slope angles (see Figure 4), with stresses transferred to the mid-plane of the GCLs in the various cross sections, see Figure 5. The cross sections replicate landfill cover scenarios and include geomembranes covering the GCLs and the GCLs used as a single barrier material. Stresses have been transferred to the mid-plane of the GCLs as of March, 1995 by cutting all geosynthetics, with the exception of the lower geotextiles or geomembranes of the respective GCLs. The lower materials are left in the anchor trenches at the top of the slopes. All mid-plane test sections have been stable to date. Full details are available from an EPA workshop on the subject, see Daniel, et al. [10].

CONCLUSIONS AND RECOMMENDATIONS

This paper on GCL perspectives has attempted to provide a link between the origins of GCLs and the current ASTM symposium on testing issues. In this latter regard, hydraulic conductivity and shear strength are the focus of many of the current testing efforts. This is altogether proper since GCLs function as hydraulic barriers and, as such, must be stable in many geometric configurations including stability on relatively steep slopes. With proper test methods and procedures it is likely that representative tests will be developed and implemented accordingly.

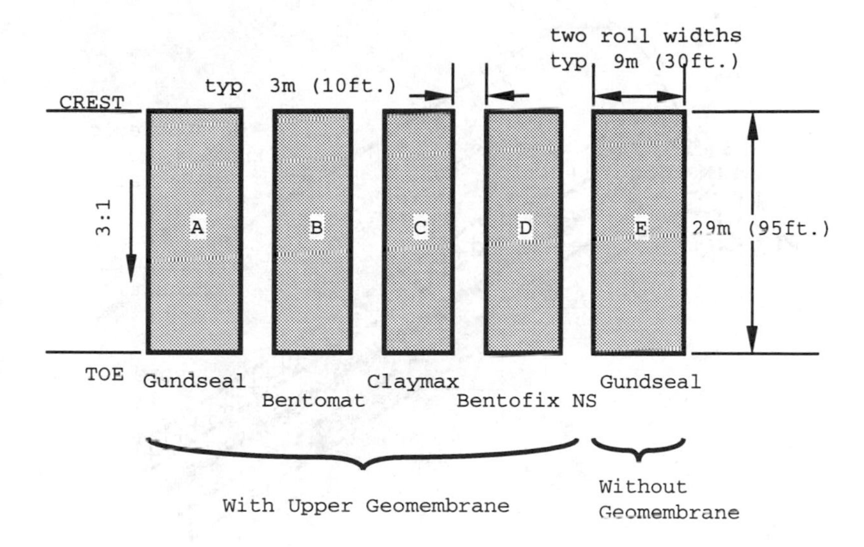

(a) Dimensions of 3 to 1 slopes

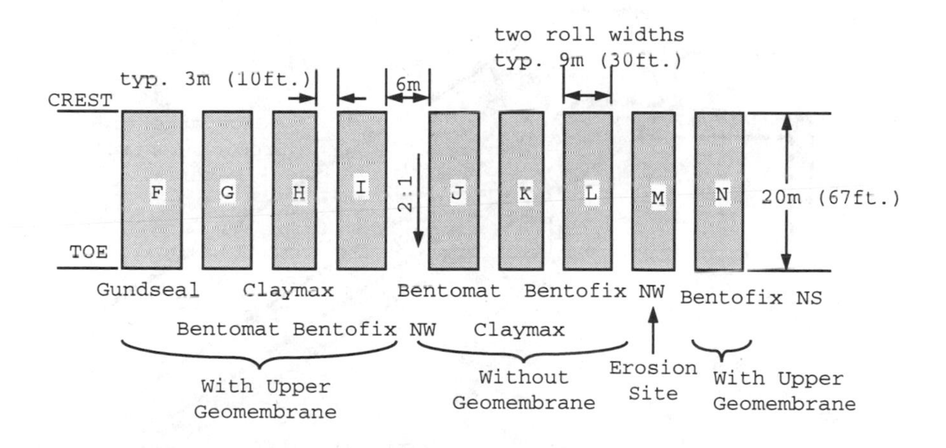

(b) Dimensions of 2 to 1 slopes

FIG. 4--Plan view of EPA-GCL slopes in Cincinnati, Ohio

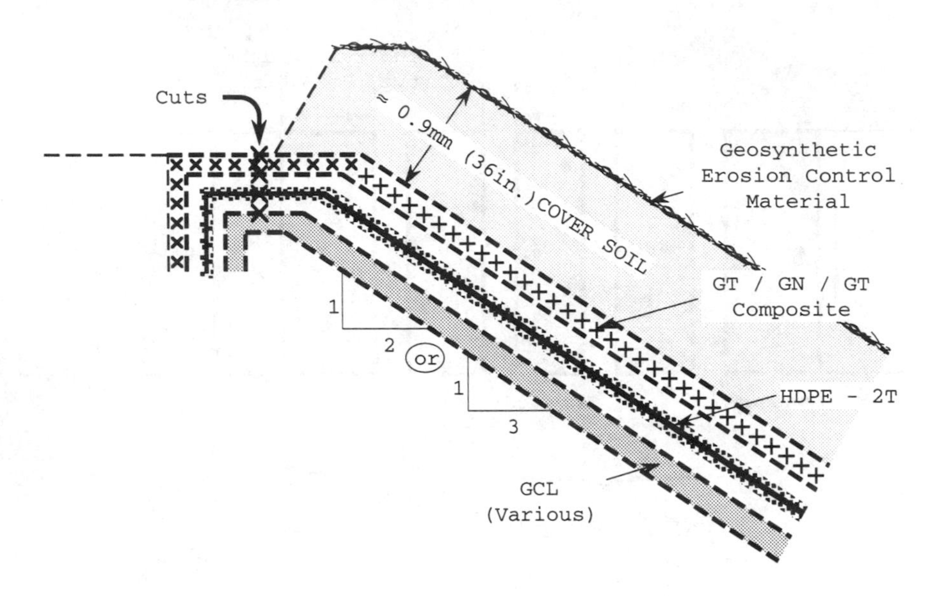

WITH GEOMEMBRANE

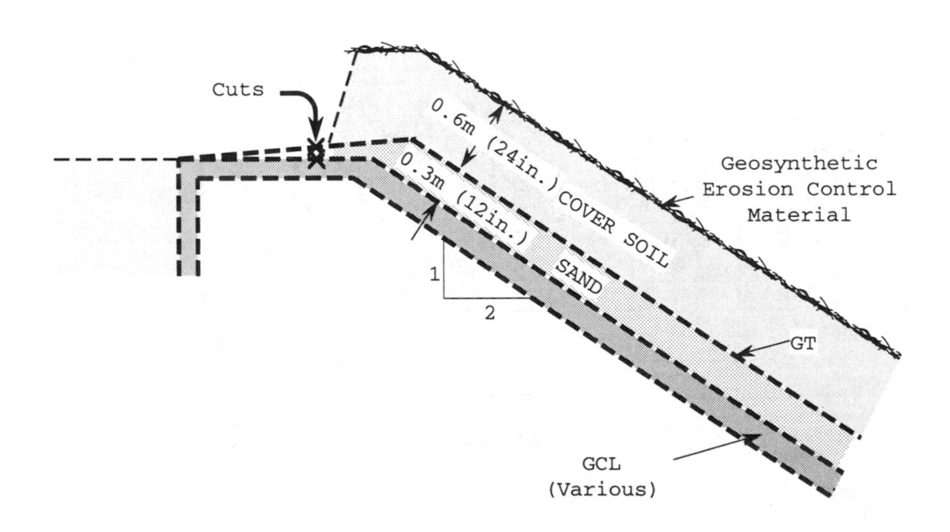

WITHOUT GEOMEMBRANE

FIG. 5--Cross sections of EPA-GCL slopes in Cincinnati, Ohio

An area where concern exists, however, is in the field deployment and backfilling of GCLs. As with all geosynthetic materials, GCLs are relatively thin. Rough subgrades and/or coarse cover soils can compromise their thickness leading to reduced (or perhaps even unacceptable) performance. While some of the more rigorous specifications call for maximum particle sizes of 12 mm (0.5 in.) on the subgrade or in the cover soil, this topic is by no means settled. In this regard, the recommendation of constructing site specific field test pads is offered. Such test pads should use the simulated soil subgrade, candidate GCL product and anticipated soil cover material. Trafficking with the anticipated maximum weight vehicles under different cover soil thicknesses and then careful exhuming of the GCL should provide proper insight into the minimum allowable cover soil thickness for a given soil and site. Test pads of this type are highly recommended. Even further, using the ASTM D5818 guide on installation procedures of test pad construction it would be a worthwhile near-term goal to develop a procedure for assessing the extent of possible damage on the exhumed GCLs.

ACKNOWLEDGMENTS

The financial support of the member organizations of the Geosynthetic Research Institute is greatly acknowledged.

REFERENCES

[1] Madsen, F. and Nüesch, R., "Characteristics and Sealing Effects of Bentonites," in Geosynthetic Clay Liners, A. A. Balkema, Rotterdam/Brookfield, 1995, pp. 31-50.

[2] Egloffstein, T., "Properties and Test Methods to Assess Bentonites Used in Geosynthetic Clay Liners," in Geosynthetic Clay Liners, A. A. Balkema, Rotterdam/Brookfield, 1995, pp. 51-72.

[3] Trauger, R. J., "The Structure, Properties and Analysis of Bentonite in Geosynthetic Clay Liners," Proc. GRI-8 Conference on Geosynthetic Resins, Formulations and Manufacturing, IFAI, St. Paul, MN, 1995, pp. 205-218.

[4] Koerner, R. M., Designing with Geosynthetics, 3rd Edition, Prentice Hall Publ. Co., Englewood Cliffs, NJ, 1994, 783 pgs.

[5] Fuller, J. M., "Landfill Cap Designs Using Geosynthetic Clay Liners," in Geosynthetic Clay Liners, A. A. Balkema, Rotterdam/Brookfield, 1995, pp. 129-140.

[6] Erickson, R. B. and Anderson, J. D., "The Manufacturing and Application of a Geomembrane-Supported Geosynthetic Clay Liner," Proc. GRI-8 Conference on Geosynthetic Resins, Formulations and Manufacturing, IFAI, St. Paul, MN, 1995, pp. 235-242.

[7] von Maubeuge, K. P. and Heerten, G., "Needle Punched Geosynthetic Clay Liners," Proc. GRI-8 Conference on Geosynthetic Resins, Formulations and Manufacturing, IFAI, St. Paul, MN, 1995, pp. 219-228.

[8] Daniel, D. E. and Boardman, B. T., "Report of Workshop on Geosynthetic Clay Liners," EPA/600/R-93/171, U.S. Environmental

Protection Agency, Cincinnati, OH, 1993, 106 pgs.
[9] Daniel, D. E. and Estornell, P. M., "Compilation of Information on Alternative Barriers for Liner and Cover Systems," EPA/600/2-91/002, U.S. Environmental Protection Agency, 1991, 87 pgs.
[10] Daniel, D. E. and Scranton, H. B., Proc. of 1995 Workshop on Geosynthetic Clay Liners (in press).
[11] Koerner, R. M., Gartung, E. and Zanzinger, H., Geosynthetic Clay Liners, A. A. Balkema, Rotterdam/Brookfield, 1995, 245 pgs.
[12] Well, L. W. and von Maubeuge, K. P., Symposium on Testing and Acceptance Criteria for Geosynthetic Clay Liners, STP-1308, ASTM, Philadelphia, PA, 1996.
[13] Scheu, C., Johannssen, K. and Saathoff, F., "Nonwoven Bentonite Fabrics - A New Fiber Reinforced Mineral Liner System," 4th Intl. Conf. on Geosynthetics, G. den Hoedt, Ed., A. A. Balkema, Rotterdam, 1990, pp. 467-472.
[14] Rathmayer, H., "Guidelines on the Use of Liners in Highway Construction," in Geosynthetic Clay Liners, A. A. Balkema, Rotterdam/Brookfield, 1995, pp. 113-128.
[15] Schmidt, R., "GCL Installation in a Water Protection Area for the A96 Motorway Near Leutkirch, Germany," in Geosynthetic Clay Liners, A. A. Balkema, Rotterdam/Brookfield, 1995, pp. 199-206.
[16] Rau, M. and Dresher, J., "Measurement and Control System for the Upper Basin of the Reisach-Rabenleite Pumped Storage Power Station," in Geosynthetic Clay Liners, A. A. Balkema, Rotterdam/Brookfield, 1995, pp. 207-218.
[17] Heerten, G. and List, F., "Rehabilitation of Old Liner Systems in Canals," 4th Intl. Conf. on Geosynthetics, G. den Hoedt, Ed., A. A. Balkema, Rotterdam, 1990, pp. 453-456.
[18] Koerner, G. R., Eberlé, M. and Koerner, R. M., "Potential Benefits of GCLs as Canal Liners," Geotechnical Fabrics Report, Vol. 13, No. 1, IFAI, January, 1995, pp. 36-39.
[19] Heerten, G., Saaathoff, F., von Maubeuge, K. P. and Hahn, S., "Geosynthetic Clay Liners as Gas and Liquid Barriers in Caps and Polluted Areas," Proc. 2nd Intl. Conf. on Soil Contamination, Vitoria Gasteiz, IHOBE Publications, Balboa, Spain, 1994, pp. 1-19.
[20] Heerten, G., von Maubeuge, K. P., Simpson, M. and Mills, C., "Manufacturing Quality Control of Geosynthetic Clay Liners - A Manufacturers Perspective," Proc. GRI-6 Conference on MQC/MQA and CQC/CQA of Geosynthetics, R. M. Koerner, and Y. G. Hsuan, Eds., IFAI, St. Paul, MN, 1993, pp. 86-95.
[21] Weiss, W., Siegmund, M. and Alexiew, D., "Field Performance of a Geosynthetic Clay Liner Capping System Under Simulated Waste Subsidence," Proc. Geosynthetics '95, Altanta, GA, IFAI, 1995, pp. 641-654.
[22] GeoSyntec Inc., study in progress for U.S. EPA.

Special Testing Procedures

Dhani B. Narejo[1] and Yun Zhou[2]

EFFECT OF MOISTURE CONTENT ON FREE SWELL OF THE CLAY COMPONENT OF GEOSYNTHETIC CLAY LINERS

REFERENCE: Narejo, D. B. and Zhou, Y., **"Effect of Moisture Content on Free Swell of the Clay Component of Geosynthetic Clay Liners,"** *Testing and Acceptance Criteria for Geosynthetic Clay Liners, ASTM STP 1308,* L. W. Well, Ed., American Society of Testing and Materials, 1997.

ABSTRACT: This paper discusses the effect of initial moisture content on free swell of the clay component of Geosynthetic Clay Liners (GCLs). Geosynthetic Research Institute (GRI) test method GCL-1 was used to perform all free swell tests included in this paper. Three commercially available GCLs were used in the study. For each GCL, free swell tests were performed at various initial moisture contents. For two GCLs, the free swell was seen to decrease linearly with an increase in the initial moisture content of the test specimens. For the third GCL, the free swell was found to decrease non-linearly with an increase in the initial moisture content. A procedure for correcting the free swell for the initial moisture content of GCLs is suggested. Using this procedure, the free swell of the clay component of a GCL test specimen at a desired initial moisture content can be calculated from the experimentally determined free swell at a different initial moisture content.

KEYWORDS: geosynthetic clay liners, free swell

Geosynthetic Research Institute (GRI) test method GCL-1 [1] is being used increasingly for measuring free swell of the clay component of Geosynthetic Clay Liners (GCLs). The test method is intended to be used primarily for quality control / quality assurance (QC/QA) and conformance testing purposes. As such, the goal of the procedure is to identify changes or variations in the quality of clay (such as, the type and percentage of clay minerals) during manufacturing and installation of GCLs.

The magnitude of free swell of a given type of clay depends, among other factors, on the initial moisture content of the free swell test specimen [1]. Clay can absorb or lose water depending on its moisture content and the humidity in the surrounding atmosphere.

---

[1]Research Associate, Department of Civil Engineering, Carleton University, 1125 Colonel By Drive, Ottawa, Ontario K1S 5B6

[2]Civil Engineer, IT Corporation, 2790 Mosside Blvd., Monroeville, Pittsburgh, PA 15146.

Therefore, a change in free swell of the clay component of a GCL test specimen may be the result of the change in moisture content rather than the quality of clay. The need to assess and isolate the effect of initial moisture content on free swell of the clay component of GCLs is obvious. Once the relationship between moiture content and free swell is understood, changes in the quality of clay component of GCLs can be more easily identified using GRI test method GCL-1.

TEST PROCEDURE

GRI test method GCL-1 was used to perform all free swell tests included in this paper. The test procedure uses the California Bearing Ratio (CBR) test equipment [2] to perform the test. The CBR equipment consists of the CBR mold with an extension collar, porous base plate, porous top plate with an extension rod and a tripod with dial gage. The CBR mold is 150 mm in diameter and 175 mm long. The extension collar of the mold is the same diameter but 50 mm long.

The test method requires the separation of clay, including any adhesives, fibers, etc., from a candidate GCL. The clay and the associated material is passed through a No. 4 U. S. standard sieve. The CBR mold is placed in a container and levelled so that it is horizontal. One hundred (100) grams of the sieved portion of clay is placed in the CBR mold sandwiched between two filter papers. The clay is compacted by a circular and tamping action of the porous plate and is, therefore, in a comparatively loose state. The porous cover plate with the extension rod is placed on top of the test specimen. No surcharge is applied except the weight of the porous cover plate and the accompanying extension rod. The tripod is placed around the CBR mold assembly and the dial gage stem is brought in contact with the adjustable stem of the top porous plate. After the initial dial gage reading is recorded, de-ionized and de-mineralized water is added to the CBR mold and the container housing the CBR mold assembly. Dial gage readings are taken over a period of 24 hours. Dial gage reading at the end of 24 hours minus the initial dial gage reading is regarded as the free swell of the clay component of the GCL.

The moisture content of the free swell test specimens was determined according to ASTM test method D-2216 [3].

TEST RESULTS

Free swell tests were performed on three commercially available GCLs at various initial moisture contents. These GCLs are identified in this paper as GCL No. 1, GCL No. 2 and GCL No. 3. The test results for GCL No. 1, GCL No. 2 and GCL No. 3 are presented in figures 1, 2 and 3, respectively. Each free swell value plotted on the figures represents a single test. The accuracy of trends shown in the figures can be improved by repeating each test more than once and then plotting average free swell against initial moisture content.

The as-received moisture content of GCL No. 1 was around 30 % whereas that of GCL No. 2 and GCL No. 3 was around 8 %. The lower-than-as-received moisture contents plotted on the figures were obtained by drying the clay component of test specimens in an oven at 105 ^{o} C for the necessary time period. The higher than as-received moisture contents were obtained by adding the required amount of de-ionized water to the clay component of test specimen and equilibrating for 24 hours. Figures 1 and 2 show that the free swell of GCL No. 1 and GCL No. 2 decreases approximately linearly with an increase in the initial moisture content of the test specimens. Figure 3 shows that the free swell of GCL No. 3 decreases non-linearly with increase in the initial moisture content of

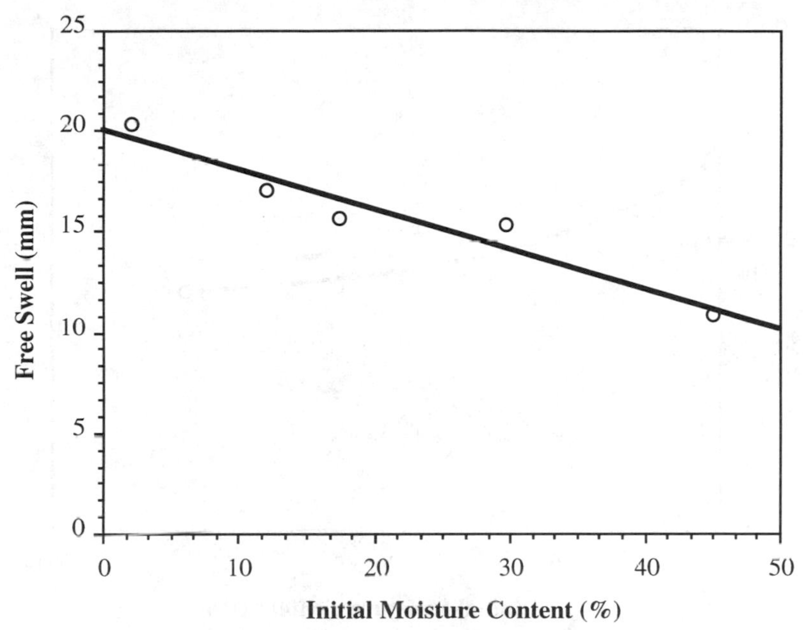

Figure 1-- Effect of Initial Moisture Content on Free Swell of GCL No. 1

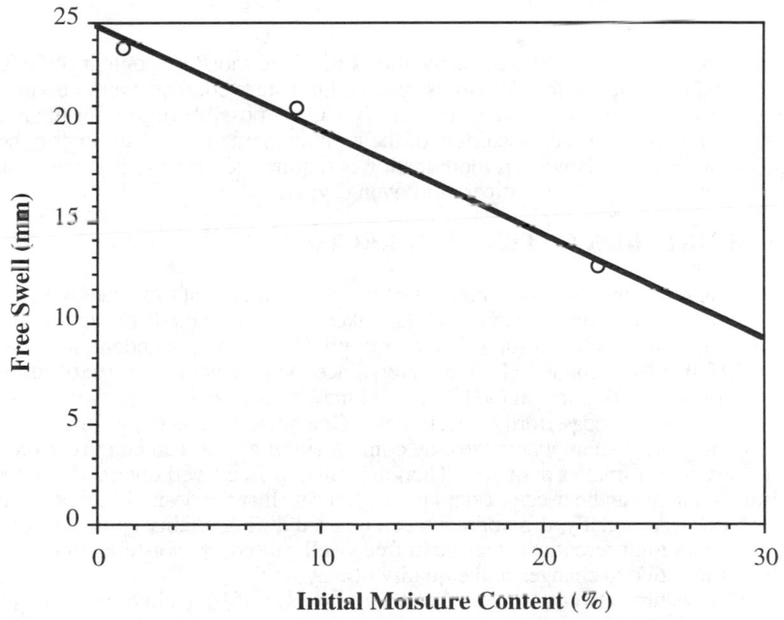

Figure 2-- Effect of Initial Moisture Content on Free Swell of GCL No. 2

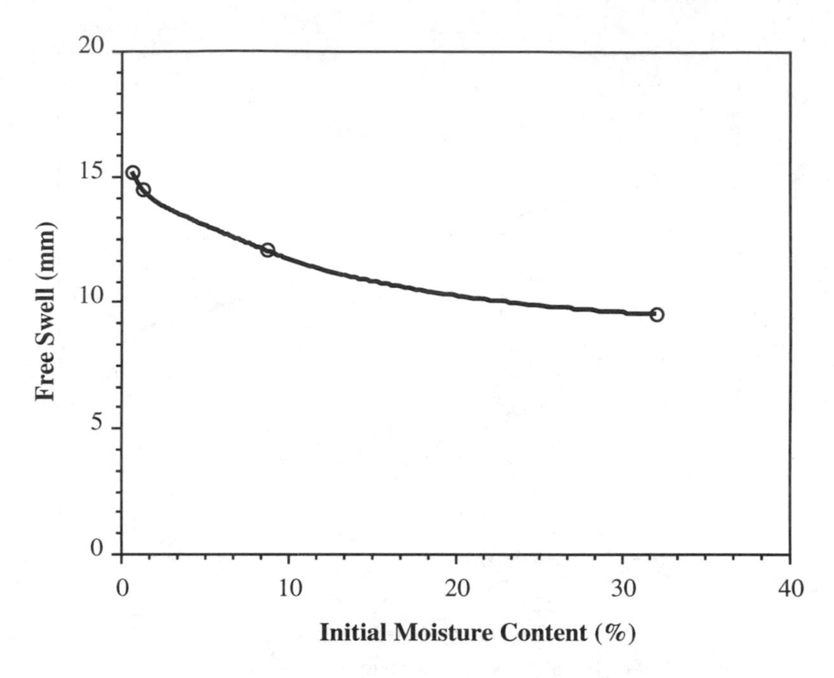

Figure 3-- Effect of Initial Moisture Content on Free Swell of GCL No. 3

the test specimens. As mentioned previously, the as-received moisture content of GCL No. 3 was around 8 % and the lower-than-as-received moisture contents were obtained by drying the free swell test specimens in an oven at 105°C. A possible reason for the non-linear relationship may be the adverse effect of the high temperature oven drying on the clay component of GCL No. 3. However, more testing is required to prove that the non-linear relationship is caused by the high temperature oven drying.

INITIAL MOISTURE CONTENT CORRECTION

The as-manufactured moisture content of the clay component of some GCLs can be quite high. At the higher moisture contents, clay takes the form of clods or lumps. It may not be possible to pass the clay in such a state through a No. 4 U. S. standard sieve as required by GRI Test Method GCL-1. Moreover, since the clay is in the form of lumps, an even placment of only 100 grams in the CBR mold may be considerably difficult. As such, the free swell test may not be performed accurately. One alternative is to perform the test at a lower initial moisture content where the clay component of a GCL can be pulverized or, at least, broken down into smaller particles. Then the value of free swell obtained at a lower initial moisture content can be used to calculate the free swell at the desired higher initial moisture content. Additionally, this approach can be used during quality control testing in a manufacturing plant to differentiate changes in free swell caused by moisture content variation from those due to changes in the quality of clay.

The relationship between weight of water and weight of solids in a soil test specimen can be expressed as follows:

$$M_c = \frac{W_w}{W_s} \tag{1}$$

where

M_c = moisture content (%)
W_w = weight of water (grams), and
W_s = weight of solids (grams)

Also, the weight of water and weight of solids are related to the total weight of soil test specimen as follows:

$$W = W_w + W_s \tag{2}$$

where

W = total weight of a test specimen = 100 grams for GRI test method GCL-1.

Combining equations (1) and (2), one obtains;

$$W_s = \frac{W}{1+M_c} \tag{3}$$

Equation (3) shows that for a constant total weight, W, the weight of solids depends only on the moisture content of the test specimen. Thus, Equation (3) can be used to convert abscissae in figures 1, 2, and 3 to weight of solids rather than moisture content. It can be shown that plots between free swell and weight of solids for GCL No. 1 and GCL No. 2 are linear and that for GCL No. 3 non-linear. Therefore, for a given type of GCL with a linear relationship between weight of solids and free swell, the following relationship holds:

$$F = C\,W_s \tag{4}$$

where

F = free swell (mm)
C = constant (mm/grams)

Substituting the value of W_s from Equation (3) in Equation (4), one obtains;

$$F = \frac{C\,W}{1+M_c} \tag{5}$$

Assuming that a free swell test is performed at an initial moisture content M_{c1}, Equation (5) can be written as follows:

$$F_1 = \frac{C\,W}{1+M_{c1}} \tag{6}$$

However, one may want to know the free swell of the clay component of the same GCL at a different initial moisture content, say M_{c2}. Then, Equation 5 can be written as follows:

$$F_2 = \frac{C\,W}{1+M_{c2}} \tag{7}$$

Dividing Equation (7) by Equation (8), one obtains;

$$\frac{F_2}{F_1} = \frac{1+M_{c1}}{1+M_{c2}} \quad (8)$$

or

$$F_2 = F_1\left(\frac{1+M_{c1}}{1+M_{c2}}\right) \quad (9)$$

Using Equation (9), free swell, F_2, of clay component of a GCL can be calculated at a desired initial moisture content, M_{c2}, from an experimentally determined free swell, F_1, at an initial moisture content M_{c1}.

Example: For GCL No. 1 in Figure 1, free swell at a moisture content of 2 % was experimentally determined to be about 20 mm. What would be the free swell for this GCL at a moisture content of 15 % ?

Moisture content = M_{c1} = 2 % = 0.02
Free swell = F_1 = 20 mm
Moisture content = M_{c2} = 15 % = 0.15
Free swell = F_2 = ?

Using Equation (9),

$$F_2 = 20\left(\frac{1+0.02}{1+0.15}\right)$$

or

$$F_2 = 17.7 \text{ mm}$$

Comparing the above calculated value of F_2 with the experimentally determined values of free swell in Figure 1, it can be seen that the moisture content correction approach yields fairly accurate values of free swell for GCL No. 1. The same can also be shown for GCL No. 2 and less than 8 % moisture content for GCL No. 3.

CONCLUSIONS

Based on the limited testing performed for this paper, the following conclusions can be drawn:

(1) The free swell of clay component of GCLs decreases approximately linearly with an increase in the initial moisture content.

(2) The free swell of clay component of some GCLs may be affected by drying in oven at 105 ^{o} C.

(3) The initial moisture content correction procedure given in this paper can be used to calculate free swell at a desired initial moisture content from the experimentally determined free swell at a different initial moisture content.

RECOMMENDATIONS

Free swell test specimens should be protected from loss or gain in moisture. Moisture content test [3] should be performed on the clay component of a GCL before the free swell test. Should the moisture content of the clay component be different, water may need to be added or the clay may need to be dried. Manufacturers of GCLs should be consulted for the safe drying temperatures. The moisture content correction procedure given in the paper can be used to calculate free swell at a desired initial moisture content from the experimentally determined free swell values at a different initial moisture content. Further tests should be conducted to verify the findings in this paper.

REFERENCES

[1] GRI Test Method GCL-1, "Standard Test Method for Swell Measurement of the Clay Component of GCLs".

[2] ASTM Test Method D-1883, "Standard Test Method for CBR (California Bearing Ratio) of Laboratory Compacted Soils".

[3] ASTM Test Method D-2216, "Laboratory Determination of Water (Moisture) Content of Soil, Rock and Soil-Aggregate Mixtures".

Timothy D. Stark[1]

EFFECT OF SWELL PRESSURE ON GCL COVER STABILITY

REFERENCE: Stark, T. D., **"Effect of Swell Pressure on GCL Cover Stability,"** *Testing and Acceptance Criteria for Geosynthetic Clay Liners, ASTM STP 1308,* Larry W. Well, Ed., American Society for Testing and Materials, 1997.

ABSTRACT: This paper describes the importance of bentonite swell pressure on the stability of cover systems that incorporate a geosynthetic clay liner (GCL). The results of a one-dimensional swell test indicate that the field swell pressure of a needle-punched GCL ranges from 35 to 40 kPa. An effective normal stress at or near this swell pressure may be required to maximize the contact area between the GCL and geomembrane and increase the static and seismic stability of a GCL cover. Since an effective normal stress of 35 to 40 kPa is probably not practical and a soil cover is usually not immediately placed, it is recommended that free swell conditions be assumed for GCL shear testing and the slope be designed using the resulting shear strength parameters. Suggestions for modifying existing products to increase GCL cover stability are also presented.

KEYWORDS: geosynthetic clay liners, swell pressure, slope stability

For both hazardous and municipal solid waste containment facilities, the required strategy of the U.S. Environmental Protection Agency (EPA) for environmental protection is a composite liner and cover system. This composite system usually consists of a geomembrane placed in intimate contact with a compacted clay liner (CCL). Intimate contact is necessary so that any liquid passing through a hole in the geomembrane cannot spread laterally from the hole and approach the clay over a greater wetted area than the

[1] Associate Professor of Civil Engineering, University of Illinois at Urbana-Champaign, 205 N. Mathews Avenue, Urbana, IL 61801, (217) 333-7394.

hole itself. In recent years, geosynthetic clay liners (GCLs) are increasingly being chosen to replace compacted clay liners.

Some of the advantages of GCLs over CCLs are: lower and more predictable cost, prefabricated/manufactured quality, easier and faster construction, field hydraulic conductivity testing is not required, engineering properties are readily available, hydraulic conductivity is more resistant to cycles of wetting and drying, freeze/thaw cycles do not significantly increase hydraulic conductivity, the smaller thickness results in more air space, and repair is easier. Some of the disadvantages of GCLs over CCLs include smaller leachate attenuation capacity, shorter containment time, lower internal and interface shear strength, larger post-peak shear strength loss, and possibly higher long-term flux because of a reduction in bentonite thickness under the applied normal stress [1].

The first documented use of a GCL in a waste containment facility occurred in 1986 at a Waste Management of North America, Inc. site in Calumet City, Illinois [2]. The product used was manufactured by enclosing bentonite mixed with an adhesive between a woven and an open weave geotextile. This product is referred to as GCL A in this paper and was manufactured by Clem Environmental Corporation. Prior to this, bentonite blankets were primarily used for foundation waterproofing. The first of these waterproofing products was introduced in 1965 and consisted of bentonite sandwiched between corrugated cardboard panels. In 1991 the term geosynthetic clay liner was applied to bentonite blankets used in landfill liner or cover systems and other containment projects. At present, there are four main types of GCL products available. Two of these products consist of bentonite sandwiched between geotextiles that are needle-punched together. In these products the bentonite is sandwiched between a woven and nonwoven geotextile or two nonwoven geotextiles. Another product involves stitch bonding two woven geotextiles together to reinforce the bentonite. The last product consists of bentonite adhered to a high density polyethylene (HDPE) geomembrane.

The acceptance and use of GCLs in waste containment facilities has increased yearly since 1986. Market consumption of GCLs in 1995 was 54 million square feet and is estimated to increase by 8 percent to 58 million square feet in 1996 [3,4]. However, there are a number of questions regarding the design and acceptance of GCLs in waste containment facility liner and cover systems. This paper primarily addresses the design and acceptance of GCLs in landfill cover systems. In particular, the effect of bentonite swell pressure and bentonite extrusion into the geomembrane/GCL interface on the stability of GCL cover systems will be investigated.

EPA SLOPE STABILITY RESEARCH PROJECT

A GCL slope stability research project was initiated to investigate the stability of GCL cover systems. As part of this research project, four GCL manufacturers participated in the construction of fourteen landfill cover test pads. The GCL test pads were constructed in Cincinnati, Ohio at an operating waste containment facility. Nine test pads were constructed on a 2:1 slope while five test pads were constructed on a 3:1 slope. Figure 1 presents a plan view of the nine test pads constructed on the 2:1 slope. The test pads are 8 to 9 m wide and approximately 20 m long. The instrumentation allows monitoring of the geosynthetics deformation and the moisture content of the bentonite. Figure 2 presents a cross-section of Test Pads G and H. It can be seen that the GCL was placed on the natural subgrade and overlain by a 1.5 mm (60 mil) textured HDPE geomembrane (GM). The geomembrane was then overlain by a geosynthetic drainage composite (GT/GN/GT), 1 m of compacted soil, and a geosynthetic erosion mat (Figure 2).

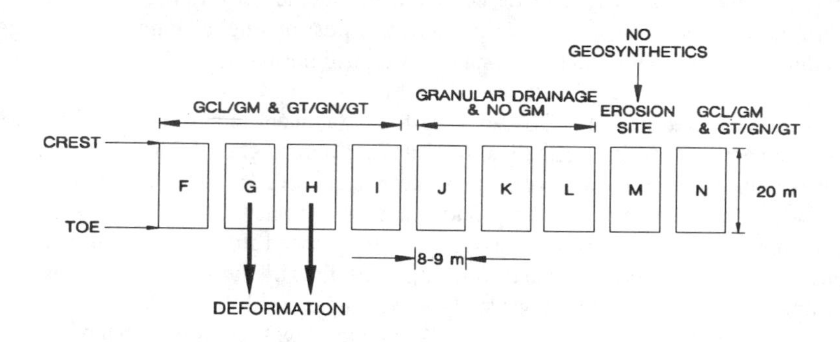

Figure 1. Plan View of GCL Test Pads on the 2:1 Slope

Approximately 20 days after installation of Test Pad H, the overlying textured geomembrane pulled out of the anchor trench and displaced 20 to 25 m to the bottom of the 2:1 slope. The GCL installed in Test Pad H consisted of two woven geotextiles stitch bonded together. The geotextiles are woven slit-film multifilament fabrics. This product is referred to as GCL B in this paper and was manufactured by Clem Environmental Corporation. In summary, deformation occurred at the interface between the woven geotextile and the overlying textured geomembrane.

Approximately 50 days after installation of Test Pad G, the overlying textured geomembrane also pulled out of the anchor trench and displaced approximately 20 meters to the bottom of the 2:1 slope. The GCL installed in Test Pad G consisted of a woven and nonwoven geotextile with needle punching providing internal support. This product is referred to as GCL C in this paper and was manufactured by Colloid Environmental Technologies Company. The deformation observed in Test Pads G and H and other field experiences, e.g., Cowland [5], clearly illustrate the importance of the GCL/geomembrane interface on the stability of landfill cover systems (Figure 2).

In Test Pads G and H, the woven geotextile of the GCL was placed in contact with the overlying textured geomembrane. A woven geotextile is used to promote intimate contact between the bentonite and geomembrane, which is required by the EPA. A woven geotextile allows bentonite to extrude through the textile during bentonite hydration and thus satisfy the requirement of intimate contact.

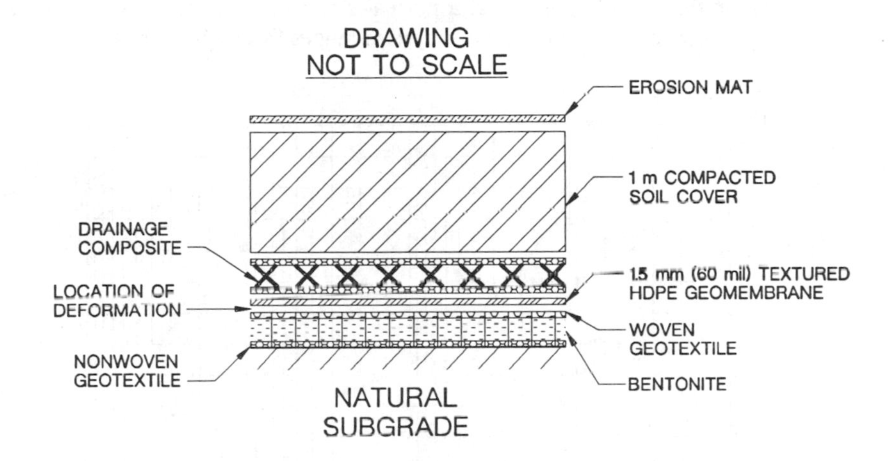

Figure 2. Cross-Section of Test Pad G

The extruded bentonite also may create a slope stability problem along the geomembrane/woven geotextile interface. The deformations observed in Test Pads G and H occurred between the woven geotextile and the textured geomembrane and are due to bentonite reducing the shear resistance of the geotextile/geomembrane interface. In summary, the requirement of intimate contact between the bentonite and geomembrane presents a dilemma for the static and seismic stability of GCL cover systems. The stability of a cover system appears to be related to the amount of bentonite that will extrude into the geomembrane/GCL interface. It is proposed herein that the amount of bentonite extrusion is controlled by the swell pressure of the bentonite or GCL. The swell pressure is defined as the normal stress required to maintain zero volume change or expansion.

FIELD SWELL PRESSURE OF GCLs

A number of researchers, e.g., [6,7], have conducted laboratory tests to estimate the swell pressure of GCLs. Leisher [6] used 0.15 m by 0.15 m square samples in dead weight and air-activated consolidometers to estimate the relationship between swell or vertical deformation and the applied normal stress. In each test the hydration fluid covered the GCL specimen and vertical deformation versus elapsed time data were recorded until equilibrium was achieved. These tests were conducted at different normal stresses to establish the relationship between vertical deformation and applied normal stress. The GCL was placed between a rigid sub-base and rigid loading platen. As a result, the full deformation or swell pressure of the GCL was measured. A typical series of swell tests for GCL A is shown in Figure 3. It can be seen that the GCL swell pressure is approximately 130 kPa. In other words, a normal stress of approximately 130 kPa is required to result in zero vertical strain. The data also suggests that more than a 150 percent increase in thickness is possible at a normal stress of approximately 7 kPa. The normal stress typically encountered in a cover system ranges from 7 to 15 kPa.

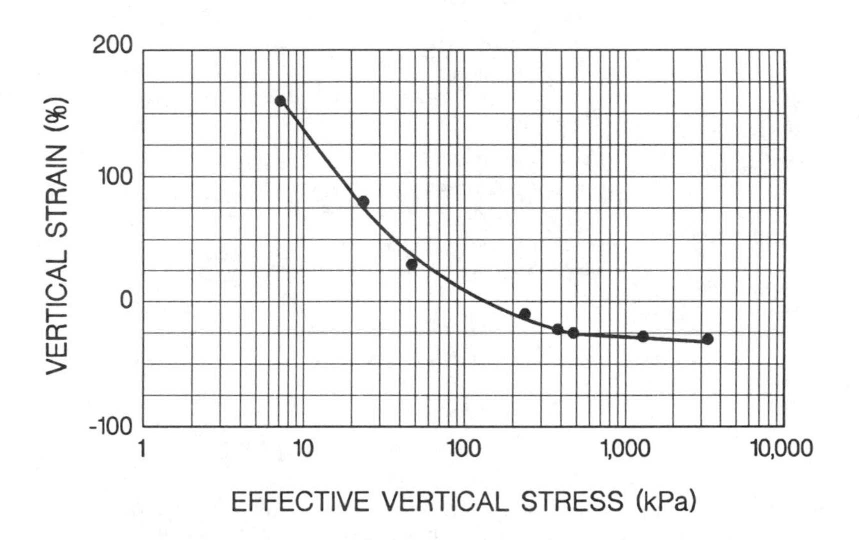

Figure 3. Swell Behavior of GCL A on a Rigid Base [7]

In a cover system, the GCL is usually placed on a lightly compacted soil layer that overlies the waste. This soil layer is usually lightly compacted because of the compressible waste underlying the compaction equipment. In addition, the overlying geomembrane is usually not in complete contact with the GCL. As a result, it was of interest to measure the swell pressure of GCLs using a flexible system to simulate field cover conditions. It was anticipated that a swell pressure less than 130 kPa would be measured because bentonite could extrude and/or deform into the nonwoven geotextile, woven geotextile/textured geomembrane interface, and granular sub-base instead of reacting against a rigid, non-porous sub-base.

To accomplish this objective, a swell pressure test was conducted in a 0.3 m by 0.3 m square consolidometer during this study. The nonwoven geotextile of the GCL was placed on a horizontal layer of sand with a unit weight of approximately 17 kN/m^3, which simulates a final interim cover layer. The consolidometer was installed in an automated INSTRON loading machine and the swell pressure was measured using a load cell. The load cell was installed between a rigid steel rod and the cross-arm of the loading machine.

The rigid steel rod was connected to the top of the loading platen and thus transmitted the swell pressure directly to the load cell. A plexiglass loading platen was placed in contact with a textured 1.5 mm (60 mil) HDPE geomembrane that was placed in contact with the upper surface (woven geotextile) of the dry, or as-received, GCL.

After placing the loading platen in contact with the geomembrane, the cross-arm of the loading machine was fixed. It should be noted that no normal stress was applied to the GCL by the loading machine prior to hydration. The cross-arm remained stationary during the swell test so that no vertical or upward movement could occur. The resulting swell pressure was measured using the load cell. The specimen was hydrated by allowing the GCL to attract water from a container that was placed along side of the consolidometer. Rubber tubing was used to connect the water container to the sand underlying the GCL. A hydraulic head was not applied to the water in the container, and thus hydration occurred due to the suction pressures induced by the bentonite.

In summary, the bentonite was free to move into the underlying nonwoven geotextile, woven geotextile/textured geomembrane interface, and cohesionless sub-base during hydration. The pressure exerted on the fixed cross-arm during swelling was measured by the load cell and assumed to be representative of the field swell pressure.

Results of Swell Test

Figure 4 presents the results of a swell test conducted during this study on GCL C, which was used in Test Pad G. It can be seen that a swell pressure of approximately 38 kPa was measured for this GCL after 9.5 days. It should be noted that the swell pressure was still increasing after 9.5 days, and thus a field swell pressure of 35 to 40 kPa is assumed throughout this paper for this GCL. A swell pressure of 35 to 40 kPa is significantly less than 130 kPa, which was measured using a rigid sub-base and loading platen as described by Leisher [6].

The fact that more than 9.5 days is required to fully hydrate a 0.3 m by 0.3 m GCL specimen has important ramifications for the laboratory shear testing of GCLs. Conventional direct shear testing usually allows 24 to 72 hours for hydration. The results in Figure 4 suggests that the specimen may not be fully hydrated with this soaking duration. However, it should be noted that the time required for complete hydration may differ depending on the hydration conditions, e.g., suction versus a positive hydraulic head.

Examination of the GCL specimen after 9.5 days of swell revealed several interesting phenomena. First, bentonite was observed to primarily extrude at the location of the needle-punching through the woven geotextile and the interface. The needle-punching separated the filaments of the woven geotextile, which provided an opening for the bentonite to extrude through. In the non needle-punched areas of the GCL, the woven geotextile remained intact and limited the amount of bentonite extrusion.

Second, the hydrated thickness of the GCL was approximately twice the original thickness of 6 to 7 mm. Thirdly, there appeared to be only localized loss or weakening of the needle-punching due to bentonite swelling. Lastly, bentonite extruded into, but not through, the underlying nonwoven geotextile.

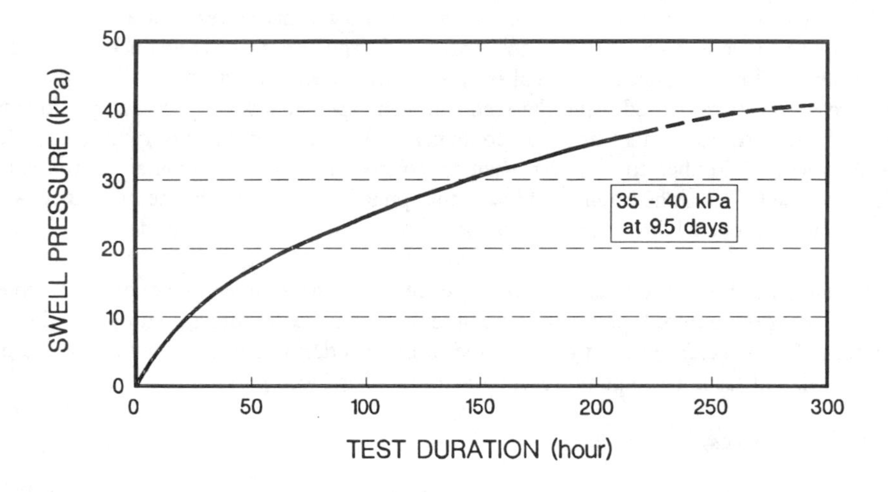

Figure 4. Swell Behavior of GCL C on a Compressible Base

In summary, the measured swell pressure for GCLs is related to the magnitude of precompression applied prior to hydration, stiffness of the test apparatus, and compliance of the simulated cover system. A swell pressure of 35 to 40 kPa appears to represent a reasonable estimate of the field swell pressure in a landfill cover that utilizes a GCL with a woven geotextile in contact with an overlying textured geomembrane. This swell pressure was measured using a deformable or flexible test configuration and no pre-compression or seating load. A field swell pressure of 35 to 40 kPa corresponds to the normal stress that is applied by 2.1 to 2.5 m of cover soil. Therefore, it can be assumed that 2.1 to 2.5 m of soil cover is needed to reduce the amount of bentonite that will extrude into the interface and increase stability. Since this soil cover thickness is usually not practical or economical, current GCL products could be modified to augment the static and seismic stability of cover systems.

Modification of Existing GCLs

One modification involves placing a nonwoven geotextile, instead of a woven geotextile, in contact with the geomembrane. This will increase the frictional resistance between the GCL and geomembrane, especially when a textured geomembrane is utilized. However, a nonwoven geotextile usually does not allow bentonite to extrude through the geotextile and develop intimate contact with the geomembrane [1]. As a result, at least one manufacturer is inserting powered bentonite in the nonwoven geotextile that is placed adjacent to the geomembrane to promote intimate contact.

In summary, the requirement of intimate contact may not be satisfied with a nonwoven geotextile of the GCL in contact with the overlying geomembrane. In addition, if bentonite cannot extrude all of the way through the nonwoven geotextile, the swell pressure may induce larger tensile forces on the internal reinforcement. This may adversely affect the long-term internal shear resistance of the GCL.

Another possible modification of existing GCLs involves eliminating the geomembrane/GCL interface from the cover system. This can be accomplished by modifying existing GCLs to incorporate the geomembrane and GCL into a single product. One way of achieving this objective is to use a co-extruded HDPE geomembrane and geonet. To create a GCL with this co-extruded geomembrane and geonet, the geonet could be filled with bentonite. The bentonite can be placed with or without an adhesive. After bentonite placement, a nonwoven geotextile can be adhered to the top of the geonet. This results in a prefabricated composite liner (PCL) system that could be used in cover or liner systems.

Other variations of the PCL include the use of an internal configuration or structure that differs from a geonet. The internal structure simply serves to facilitate bonding of the geotextile and resists the overlying normal stress, which will be discussed subsequently. A fabric encased GCL also can be obtained by bonding two geotextiles to the internal structure instead of one geotextile. This GCL could be fabricated by bonding one geotextile to the internal structure, filling the structure with bentonite, and bonding the second geotextile.

The PCL configuration results in large interface strengths between the textured geomembrane/overlying material and the nonwoven geotextile/underlying soil interface. In addition, the geonet significantly reduces the potential for internal failure or shear through the bentonite and provides some tensile resistance to the cover system. The PCL also allows the bentonite to be in intimate contact with the geomembrane.

One of the largest benefits of the geonet or internal structure is that it protects the bentonite from the effects of handling and construction and the application of normal stress after hydration. The normal stress protection is more important in a composite liner system than a cover system. Stress concentrations in a liner system can cause the hydrated bentonite to migrate to a zone of lower stress. Stress concentrations are ubiquitous in a liner system, especially around a sump and the associated plumbing features, slope transitions and benches, and geomembrane wrinkles. Anderson and Allen [1] showed that the thickness of a GCL can be significantly reduced in the vicinity of a geomembrane wrinkle. A normal stress of 240 kPa was applied to a hydrated GCL in the presence of a geomembrane wrinkle. The one-dimensional compression test showed that the bentonite migrated toward the area or void under a geomembrane wrinkle. The thickness of the

GCL under the wrinkle was 20 to 25 mm while the thickness of the GCL farthest away from the wrinkle was less than 2.5 mm. The initial thickness of the GCL was 7.5 mm. These tests were conducted using the 0.3 m by 0.3 m square consolidometer described previously. A reduced bentonite thickness can adversely affect the calculation of hydraulic equivalence between a GCL and CCL. This is of particular importance in a sump area where leachate is designed to collect.

PCL Installation and Cost

The PCL could be installed using the scheme depicted in Figure 5. It can be seen that the internal structure is not extruded to the edge of the geomembrane sheet. This allows the geomembrane from each panel to be welded together to create a continuous geomembrane barrier. In addition, granular bentonite or a strip of the proposed fabric encased GCL (two geotextiles bonded to an internal structure) could be placed in the seam area to complete the clay barrier. Another construction or seaming technique could involve placing granular bentonite on the underlying geomembrane and simply overlapping the edge of the geomembranes. This overlapping technique is currently used for other GCL products. It should be noted again that the geomembrane and internal structure are co-extruded, and thus there is no interface or weakness between the bentonite and the geomembrane.

Other applications of the PCL involve placing the geomembrane of the PCL in contact with the subgrade material and placing a geomembrane above the PCL to encapsulate the bentonite. Another application could involve not bonding a nonwoven geotextile to the top of the geonet and placing the geonet in contact with the subgrade. Preliminary testing shows that the geonet can embed into the subgrade material, which results in a large interface strength. The thickness of the geonet not embedded in the subgrade reduces the potential for hydrated bentonite to migrate due to the overlying normal stress.

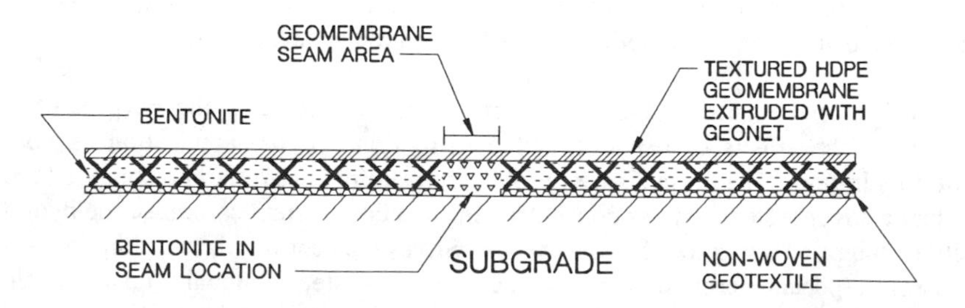

Figure 5. Possible Installation of Prefabricated Composite Liner

The main disadvantage of the PCL is cost. The manufactured cost of this GCL has not been established but it will be greater than existing GCLs. However, the PCL incorporates a geomembrane and saves a step in the installation process. As a result, the overall cost may be similar or less than an existing composite system consisting of a geomembrane and GCL. One manufacturer is currently working on fabricating prototypes of the PCL, which will aid in cost estimating.

GCL Cover Installation/Design Considerations

The laboratory data presented previously and the field performance of GCL Test Pads G and H were used to develop recommendations for the installation/design of GCLs. Clearly, the time between GCL placement and normal stress application should be minimized. If the GCL hydrates without the applied normal stress, more bentonite will likely extrude into the GCL/geomembrane interface and reduce the contact area between the GCL and overlying geomembrane.

For stability purposes, the contact area between the bentonite and geomembrane should be minimized. To minimize this contact area, a normal stress greater than or equal to the swell pressure should be applied. The swell tests conducted using a flexible system suggest that a normal stress of approximately 35 to 40 kPa needs to be applied, which corresponds to a soil thickness of 2.1 to 2.5 m. Clearly this thickness of soil cover is not practical. As a result, it is recommended that designers assume that the GCL will hydrate under free swell conditions and the slope be designed using the resulting shear strength parameters. This involves conducting laboratory shear tests that allow soaking with zero confinement until vertical expansion is completed or until the actual time between GCL placement and soil cover completion has elapsed. After vertical expansion or swell has been completed or the estimated time between GCL placement and soil cover completion has elapsed, the normal stress at which shearing will occur should be applied. Shearing can begin as soon as the GCL specimen consolidates to the applied normal stress. The resulting shear strength parameters should be used to evaluate cover stability.

Conducting GCL shear tests after completion of free swell, or the actual time that will elapse between GCL placement and soil cover construction, will provide shear strength parameters that are representative of field hydration conditions. The basis for this recommendation is that a typical GCL cover construction procedure involves placing one to two acres of GCL and geomembrane per day. As a result, it may require several days to a couple of weeks for completion of the soil cover due to geomembrane seaming, seam testing and repair, drainage layer placement, and cover soil compaction. In the interim the bentonite may hydrate due to (1) high suction pressures in the bentonite attracting moisture, (2) heat absorption by the overlying geomembrane causing subgrade moisture to migrate to the GCL, and/or (3) the GCL retaining saturated landfill gas..

REPORTING GCL FRICTION ANGLES

Since this symposium also addresses the laboratory testing of GCLs, the remainder of this paper provides recommendations for the reporting of GCL shear test results. Geosynthetic/geosynthetic and soil/geosynthetic interfaces usually exhibit stress dependent behavior. Previous testing of geosynthetic interfaces typically encountered in waste containment facilities [8,9] has shown that the stress dependent behavior results in nonlinear failure envelopes (Figure 6). The entire nonlinear failure envelope cannot be represented by a single value of friction angle. More importantly, this suggests that a friction angle applicable to a cover stability analysis is probably not applicable to a liner stability analysis.

Therefore, it has been recommended that the entire nonlinear failure envelope or a friction angle that corresponds to the average effective normal stress on the critical slip surface be used in a stability analysis [8,10]. For simplicity, most laboratories and designers report or utilize a single value of friction angle to represent the shear resistance. As a result, it is recommended that the reported friction angle utilizes a subscript to identify the range of normal stress over which the friction angle should be used in stability analyses and/or the range of normal stress at which the shear testing was conducted. The proposed notation for the peak friction angle is:

$$\phi_{1000\text{ - }6000\text{ psf}} \cong x \text{ degrees}$$

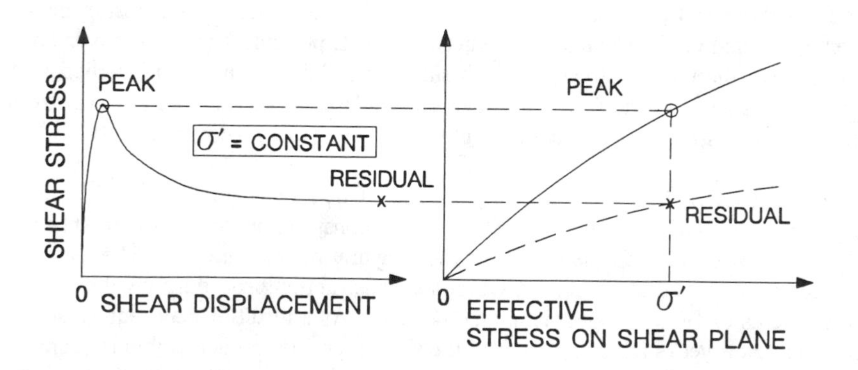

Figure 6. Shear Characteristics of Geosynthetic Interfaces

Post-Peak Friction Angles

In general, geosynthetic interfaces exhibit a peak shear strength at a shear displacement of 2 to 8 mm [8,9]. Continued displacement after the peak interface strength results in a decrease in the measured shear resistance (Figure 6). After considerable continuous shear displacement in one direction a constant minimum, or residual, shear strength is achieved. This usually results in a stress dependent or nonlinear residual failure envelope as shown in Figure 6. The magnitude of shear displacement required to reach a residual strength condition is dependent upon the interface. However, continuous shear displacement of 500 to 750 mm has been reported for textured geomembrane/nonwoven geotextile interfaces [9]. This magnitude of shear displacement may be larger than the displacement that can be achieved in one travel of a 0.3 m by 0.3 m direct shear box. If this is the case, the friction angle calculated at the end of a 0.3 m by 0.3 m direct shear test may overestimate the residual interface strength and should not be reported as a residual value.

It should be noted that reversal of a direct shear box does not result in continuous shear displacement in one direction along an interface, and thus a reversal direct shear test may also overestimate the residual interface strength. The residual strength condition is only achieved when the shear stress-displacement relationship becomes horizontal, i.e., there is no change in the measured shear stress. La Gatta [11] recommends plotting the shear stress-horizontal displacement relationship from shear tests using the logarithm of horizontal displacement to determine if a residual strength condition is achieved. This plotting technique accentuates the slope of the shear stress-displacement curve at large deformations, allowing the horizontal portion of the curve to be clearly defined. Therefore, to ensure that a residual strength condition is reached before a ring shear or direct shear test is terminated, it is recommended that the shear stress be plotted using the logarithm of horizontal displacement. Once the shear stress becomes essentially constant on a semilogarithmic plot, the test can be stopped.

Since 0.3 m by 0.3 m direct shear tests are usually terminated at a shear displacement ranging from 25 to 100 mm, it is recommended that the magnitude of shear displacement and the applicable normal stress be reported as a subscript to the post-peak friction angle. The displacement subscript, instead of a residual subscript, should be used because direct shear tests may be terminated before a residual strength condition is achieved. The proposed notation for the post-peak friction angle is presented below:

$$\phi_{50\text{ mm},\ 2000-8000\text{ psf}} \cong y \text{ degrees}$$

If a residual interface condition is achieved in the shear test, a subscript denoting a residual strength condition can be used. One such notation is shown below:

$$\phi_{r,\ 2000-8000\text{ psf}} \cong z \text{ degrees}$$

Stark and Poeppel [8] and Cowland [5] showed that a residual interface strength is usually mobilized along a sideslope and should be used in estimating the static stability of waste containment facilities. The use of seismic deformation analyses in waste containment facility design has also accentuated the need to estimate the residual interface strength or relate the measured shear strength to the level of deformation. Clearly, the laboratory shear resistance should reflect the level of seismically induced displacement that is anticipated. Therefore, adding a displacement notation to the reported friction angle will aid engineers in determining the appropriate shear strength for static and seismic stability analyses. Another reason to specify the displacement is that the shear stress-displacement relationships may not be included in the laboratory report or may not be incorporated into the stability report. Therefore, the magnitude of shear displacement may not be conveyed to the engineer, client, and/or regulatory agency. This lack of information is especially problematic when the data is incorporated into a database and subsequently published.

SUMMARY AND CONCLUSIONS

Recent field observations demonstrate the importance of the geomembrane/GCL interface strength on the stability of landfill covers. The greater the swell pressure or the smaller the applied normal stress, the more likely that bentonite will extrude into the geomembrane/GCL interface when a woven geotextile is used as the top layer of the GCL. Swell tests conducted during this study using a flexible system indicate that the field swell pressure of a needle-punched GCL probably ranges from 35 to 40 kPa. It is recommended that a normal stress or soil cover be applied as soon as possible after GCL installation to reduce the amount of bentonite that extrudes into the interface, and thus minimize the contact area between the bentonite and geomembrane.

To reduce the amount of bentonite that extrudes into the interface, the applied normal stress should be at or above the swell pressure. Since a normal stress of 35 to 40 kPa, i.e., soil cover thickness of 2.1 to 2.5 m, may not be practical and usually several days to a couple of weeks elapse between GCL placement and completion of the soil cover, it is recommended that designers assume that the GCL will hydrate under a free swell condition. This can be simulated by conducting laboratory shear tests that allow soaking with zero confinement until vertical expansion is completed or until the actual time between GCL placement and soil cover completion has elapsed. After swelling has ceased, the desired normal stress should be applied. Shearing could begin after the specimen consolidates to the applied normal stress. The resulting shear strength parameters should be used to evaluate slope stability.

Geosynthetic/geosynthetic and soil/geosynthetic interfaces usually exhibit a stress dependent or nonlinear failure envelope. Therefore, it is recommended that the reported peak friction angle utilize a subscript to identify the range of normal stress over which the friction angle is applicable and/or the range of normal stress at which the shear testing was conducted. It is also recommended that the magnitude of shear displacement and applicable normal stress be reported as a subscript to the post-peak friction angle. The displacement subscript, instead of a residual subscript, should be used because direct shear tests may be terminated before a residual strength condition is achieved.

ACKNOWLEDGMENTS

Sam Allen of TRI/Environmental in Austin, Texas supervised the 0.3 m by 0.3 m swell test on the needle-punched GCL. The writer also acknowledges Richard Thiel for his suggestions on the laboratory shear test procedure.

REFERENCES

[1] Anderson, J.D. and Allen, S.R. (1995). "What Are the Real Design Considerations When Using a Geosynthetic Clay Liner (GCL)," Proceedings of the 9th Annual Municipal Solid Waste Management Conference, Austin, Texas.

[2] Schubert, W.R. (1987). "Bentonite Matting in Composite Lining Systems." *Proceedings of Specialty Conference Geotechnical Practice for Waste Disposal*, Geotechnical Special Publication No. 13, , American Society of Civil Engineers, New York, pp. 784-796.

[3] Jagielski, K. (1994). "United States and Canada GCL Market Update." *Geotechnical Fabrics Report*, Industrial Fabrics Association International, St. Paul, Minnesota, April/May, p. 24.

[4] Industry News (1996). "Market Report Shows Geosynthetics Made Modest Gain in 1995." *Geotechnical Fabrics Report*, Industrial Fabrics Association International, St. Paul, Minnesota, May, p. 6.

[5] Cowland, J.W. (1996). "What is the Acceptable Shear Strength of a Geosynthetic Clay Liner?" *Proceedings of Symposium on Testing and Acceptance Criteria for Geosynthetic Clay Liners*, ASTM STP 1308, American Society for Testing and Materials, Philadelphia, PA.

[6] Leisher, P.J. (1992). "Hydration and Shear Strength Behavior of Geosynthetic Clay Liners." thesis submitted in partial fulfillment of the requirements for the Masters of Science Degree, Drexel University, Philadelphia, PA, 130 pp.

[7] Shan, H.Y. (1990). "Laboratory Tests on Bentonitic Blankets." thesis submitted in partial fulfillment of the requirements for the Masters of Science Degree, University of Texas, Austin, TX, 84 pp.

[8] Stark, T.D. and Poeppel, A.R. (1994) "Landfill Liner Interface Strengths From Torsional-Ring-Shear Tests," *Journal of Geotechnical Engineering*, American Society of Civil Engineers, Vol 120, No. 3, pp. 597-615.

[9] Stark, T.D., Williamson, T. A., and Eid, H.T. (1996) "Geomembrane/Geotextile Interface Shear Strength," *Journal of Geotechnical Engineering*, American Society of Civil Engineers, Vol. 122, No. 3, pp. 197-203.

[10] Stark, T.D. and Eid, H.T. (1992) "Drained Residual Strength of Cohesive Soils," *Journal of Geotechnical Engineering*, American Society of Civil Engineers, Vol. 120, No. 3, pp. 597-615.

[11] La Gatta, D.P. (1970). "Residual Strength of Clays and Clay-Shales by Rotation Shear Tests." thesis submitted in partial fulfillment of the requirements for the Ph.D. Degree reprinted as *Harvard Soil Mechanics Series* No. 86, Harvard University Press, Cambridge, MA, pp. 204.

John R. Siebken[1], Scott Lucas[2]

A COMPARISON OF SAMPLE PREPARATION METHODOLOGY IN THE EVALUATION OF GEOSYNTHETIC CLAY LINER (GCL) HYDRAULIC CONDUCTIVITY

Reference: Siebken, J. R. and Lucas, S., **"A Comparison of Sample Preparation Methodology in the Evaluation of Geosynthetic Clay Liner (GCL) Hydraulic Conductivity,"** *Testing and Acceptance Criteria for Geosynthetic Clay Liners, ASTM STP 1308,* Larry W. Well, Ed., American Society for Testing and Materials, 1997.

Abstract: The method of preparing a single needle-punched GCL product for evaluation of hydraulic conductivity in a flexible wall permeameter was examined. The test protocol utilized for this evaluation was GRI Test Method GCL-2 Permeability of GCLs. The GCL product consisted of bentonite clay material supported by a woven and a non-woven geotextile on either side. The method preparation focused on the procedure for separating the test specimen from the larger sample and whether these methods produced difficulty in generating reliable test data. The methods examined included cutting with a razor knife, scissors, and a circular die with the perimeter of the test area under wet and dry conditions. In order to generate as much data as possible, tests were kept brief. Flow was monitored only long enough to determine whether or not preferential flow paths appeared to be present. The results appear to indicate that any of the methods involved will work. Difficulties arose not from the development of preferential flow paths around the edges of the specimens, but from the loss of bentonite from the edges during handling.

Keywords: Geosynthetic Clay Liner, Flexible Wall Permeameter, Permeability, Hydraulic Conductivity

---

1 Supervisor of Technical Services, National Seal Company, Galesburg, Illinois, USA
2 Vice-President Manufacturing, Albarrie Naue LTD., Barrie, Ontario, Canada

The evaluation of a GCL for hydraulic conductivity in a flexible wall permeameter is not new. However, problems occasionally arise in the interpretation of the results, the test may show a higher flow than expected. If there were no obvious problems with the specimen, then this anomaly is sometimes blamed on "sidewall leakage". To minimize the potential for this to occur, the GRI GCL-2 Method Permeability of GCLs makes suggestions for preparing the specimen. Outside of unpublished discussions in ASTM technical committee meetings and a brief reference in a technical paper [1], the authors are not aware of anyone having looked specifically into this area of the test with respect to GCLs.

This paper reports the results of tests that were designed to examine several specimen preparation conditions to determine if any one was significantly better than the other. Due to the limited availability of test devices, each test was limited to a flow time duration of no more than five days.

TEST DEVICE, MATERIALS and PROCEDURE

There were two brands of devices involved with this study;

Device A - manufactured by Trautwein Soil Testing Equipment [2], and,
Device B - manufactured by Geotest Instrument Corporation [3].

Both devices were similar in that they were flexible wall permeameters meeting the specifications set forth in ASTM D 5084 Test Method for Measurement of Hydraulic Conductivity of Saturated Porous Materials Using a Flexible wall Permeameter.

The devices differed from each other in the diameter of the test area and in the manner in which flow was measured. Device A utilized specimens with a diameter of 0.07 m, and device B utilized specimens with a diameter of 0.10 m. To monitor flow, device A used a burette panel while device B used a piston/cylinder interface of known dimensions and a dial indicator.

The material used in this evaluation was a needle-punched GCL that consisted of a layer of bentonite clay contained by a nonwoven geotextile on one side and a woven geotextile on the other. Specimens were taken from production samples of the material. The bentonite for this product is applied to the carrier geotextile at a rate of 4.88 kg per square meter. Distilled water was used as the permeant in all tests, and in the preparation of specimens where applicable.

The general test conditions chosen for this evaluation were taken from GRI Test

Method GCL-2 Permeability of Geosynthetic Clay Liners (GCLs). Those conditions included that the test specimen be subjected to a total confining stress of 345 kPa and a back pressure of 276 kPa. The test specimen was allowed to saturate under these conditions for a minimum of 48 hours. At the end of that period, flow was initiated by raising the pressure on the influent side to 310 kPa.

The potential effect of the specimen preparation was the focus of this study and seven conditions were examined as listed in TABLE 1.

TABLE 1 --Test Specimen Preparation Conditions

Condition #1	Perimeter of Specimen Dry Geotextile Cut with Knife
Condition #2	Perimeter of Specimen Wet Geotextile Cut with Knife
Condition #3	Perimeter of Specimen Dry Geotextile Cut with Scissors
Condition #4	Perimeter of Specimen Wet Geotextile Cut with Scissors
Condition #5	Perimeter of Specimen Dry Geotextiles Cut with Circular Die
Condition #6	Perimeter of Specimen Wet Geotextiles Cut with Circular Die
Condition #7	Perimeter of Specimen Wet Geotextiles Cut with Circular Die Geotextile Edges Pressed Together

In each case, the test specimen was taken from a larger sample of at least 0.09 m^2 in area. For the specimens prepared with the perimeter dry, care was taken to spill as little bentonite from the edges as possible. However, no attempt was made to replace lost bentonite. The edges of the geotextiles were trimmed as carefully as possible, but no special attempt was made to ensure the separation of the top geotextile from the bottom geotextile.

A wet perimeter was obtained in the specimen preparation by wetting the area to be cut with distilled water from a squirt bottle and allowing the specimen to stand for a minimum of 15 minutes before attempting to separate the geotextiles. For condition #7, the edges were pressed together such that the geotextiles were in intimate contact. This was done immediately prior to loading the specimens into the permeameter.

Once loaded into the test device, each test profile consisted from top to bottom of a top cap, a soaked porous (either aluminum or ceramic) disk, a soaked filter paper disk, the test specimen, a soaked filter paper disk, a soaked porous disk, and a base pedestal. A flexible latex membrane and O-ring seals were used to separate the test profile from the confining cell water.

RESULTS

The hydraulic conductivity was calculated based upon equation (1) for a constant head test as presented in ASTM D 5084. Use of the constant head equation was justified as the hydraulic head was maintained by a pressurized system. Also, any changes in the physical level of water in the permeameters was small enough that the requirement of section 5.1.1 in ASTM D 5084 that the head loss be held constant within ±5% was met.

$$k = QL/Ath \quad (1)$$

where:

- k = hydraulic conductivity, m/s,
- Q = quantity of flow, taken as the average of inflow and outflow, m^3,
- L = length of specimen along path of flow, m,
- A = cross-sectional area of specimen, m^2,
- t = interval of time, s, over which the flow Q occurs, and
- h = difference in hydraulic head across the specimen, m of water.

The flow path length (L) was based upon the thickness of the specimens tested. For the specimens tested in device B, this was determined indirectly with a dial indicator that monitored thickness changes in the entire test profile. Changes in the specimen thickness were then calculated from a known reference point. The specimens tested in device A were calculated based upon pre and post-test thickness measurements of the specimens. The area (A) was based upon the known dimension of the porous plates used in each device. The flow (Q) was taken as the average of the inflow and the outflow. The results of the individual tests are shown in (Table 2).

TABLE 2--Individual Hydraulic Conductivity Results

Specimen Preparation	Permeability (m/s)	Elapsed Test Time[A] (days)	Total Flow[B] (m^3)	Specimen Diameter (m)
Condition 1	1.0E-11	4.97	4.19E-5	0.10
	1.2E-11	4.68	5.11E-5	0.10
	1.0E-11	1.04	2.02E-6	0.07
Condition 2	1.9E-11	0.88	2.60E-6	0.07
	7.1E-12	0.97	1.10E-6	0.07
Condition 3	9.1E-12	1.00	1.40E-6	0.07
	1.1E-10	1.00	2.19E-5	0.07
Condition 4	7.8E-11	1.00	1.36E-5	0.07
	5.8E-11	1.00	1.04E-5	0.07
	8.0E-12	1.00	1.24E-6	0.07
Condition 5	1.5E-11	4.72	5.68E-5	0.10
	1.6E-11	5.00	3.31E-5	0.10
	1.1E-11	5.65	2.20E-5	0.10
	1.1E-10	1.00	1.99E-5	0.07
	8.8E-12	2.69	3.60E-6	0.07
	8.6E-12	2.69	3.40E-6	0.07
Condition 6	1.2E-11	6.69	5.11E-5	0.10
	9.2E-12	4.20	3.67E-5	0.10
	1.1E-11	4.25	7.19E-5	0.10
	6.7E-11	3.13	3.72E-5	0.10
Condition 7	8.3E-12	7.99	3.66E-5	0.10
	3.4E-12	1.00	5.45E-7	0.07
	3.9E-12	0.96	6.75E-7	0.07
	4.2E-12	1.00	6.50E-7	0.07

Note A: Elapsed time includes only the period over which flow was monitored.
Note B: Total flow is based on the average of the recorded inflow and outflow.

These results are close to the manufacturer's stated specification of 1×10^{-11} m/s for these test conditions with only a couple of exceptions. The values for some of the specimens monitored for a longer period of time are shown graphically in Figure 1.

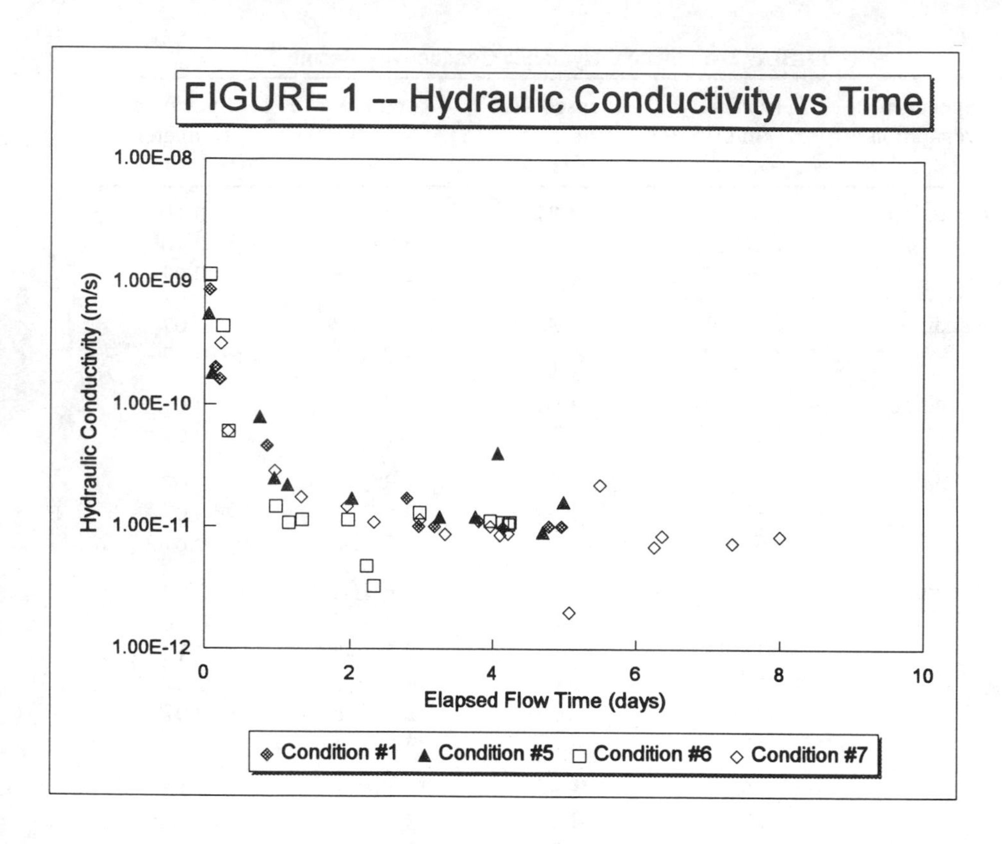

Discussion of the Results

The results show some scatter both between specimen conditions and between test devices. This is not unexpected given the number of variables involved and the short duration of the tests. However, the results as a whole are not orders of magnitude greater than the manufacturers specifications suggest they should be under the conditions examined. It was expected that the hydraulic conductivity results might be higher, due to the idea that if significant amounts of bentonite were lost from the edges of a specimen, this bentonite loss would allow the geotextile portions of the GCL to come into contact and create a preferential flow path around the edge of the specimen. The authors have seen test results on the order of 1×10^{-9} m/s where this preferential flow was thought to be the cause.

There are other points to consider, including ease of specimen preparation and the loss of bentonite from the edges during specimen preparation. It was clear during this study that the largest source of difficulty in specimen preparation was not the cutting utensil but whether the perimeter of the specimen was wet or dry. In the case of the dry specimens, it was demonstrated that the most difficult method of separating the geotextiles was with scissors. Separating the geotextiles with a razor knife was slightly less difficult, but there was still considerable loss of bentonite from the edges. The specimens prepared with the die in the dry condition fared a little better. The force of pressing the die through the layers of geotextile had the tendency to compact the geotextiles and the clay into a more solid mass. Applying water to the perimeter of the specimen area and allowing it to soak for a period of time made the preparation easier in all cases. With a wet perimeter, the three different modes of specimen preparation could be classed from most difficult to easiest as scissors, knife, and circular die.

It is considered by the authors very important to prevent the loss of bentonite from the edges of the test specimen. It was observed in the tests where the specimen had lost a large amount of bentonite that the flexible membrane would intrude significantly into the profile. This is shown schematically in Figure 2. Direct measurements of the intrusion could not be taken. It is estimated that in the most severe cases, the intrusion may have resulted in as much as a 10 to 20% reduction in the area of the test specimen. The tests with severe intrusion were terminated and restarted with fresh specimens, but it leads one to consider how much error is introduced in those tests where the diameter of the specimen is not quite equal to the diameter of the test device.

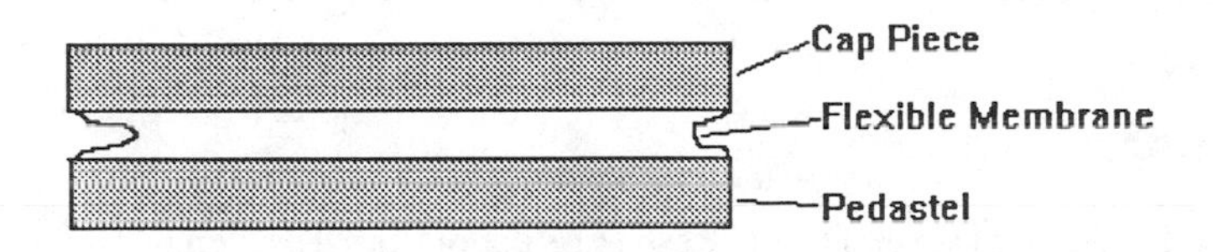

Figure 2- - Schematic of Flexible Membrane Intrusion

In condition #7 we attempted to force the geotextile portions of the GCL together. This was done with the perimeter of the specimen wet, and immediately prior to loading the specimen into the permeameter. It was hoped that this action would result in an edge through which a preferential flow path would occur and be detected by a significantly increased permeability. This was not the case. We observed during the course of the saturation period, the bentonite in the edges of the specimen appeared to swell and spread the geotextiles apart again. The inability of the effective confining stress to prevent the swelling instead appeared to allow the swelling to act as a kind of corrective measure. It remains to be seen by additional testing whether a higher effective confining stress would prevent the swelling and allow flow through the geotextile contact surfaces.

CONCLUSIONS

The following conclusions and observations are drawn from the short term permeability tests conducted with the described specimen preparation methods.

• The authors could not create a preferential flow path through the geotextiles of the GCL at the edges of the specimens.

• Of the specimen preparation methods studied, the ranking from most difficult to least difficult to perform is condition #3, #1, #5, #4, #2, and #6.

• Acceptable tests can be conducted regardless of the method of specimen preparation if sufficient care is taken in the preparation by the person conducting the test.

• The loss of bentonite from around the edges of the specimen is still considered a significant potential problem with respect to controlling the size of the GCL specimen.

• These tests focused only on the short term development of problems that might have been attributable to the method of specimen preparation. It is not known if potential problems may have surfaced later had we allowed flow to continue for longer periods of time.

• The authors recognize the need for standardization in testing. Of the methods included in this study, specimen preparation condition #6 with the perimeter wet and cut with a circular die is considered the most reproducible and desirable method to promote.

REFERENCES

[1] Shan, H. Y., and Daniel, D. E., "Results of Laboratory Tests on a Geotextile/Bentonite Liner Material", Geosynthetics 91, Industrial Fabrics Association International, St. Paul, MN, Vol. 2, pp. 517-535.

[2] Trautwein Soil Testing Equipment, P.O. Box 31429, Houston, Texas 77231.

[3] Geotest Instrument Corporation, 1840 Oak Ave., Evanston, Illinois 60201.

ACKNOWLEDGEMENTS

The authors would like to acknowledge National Seal Company and Albarrie Naue LTD for making this study possible. We would also like to acknowledge Jim Matthews of National Seal Company and Kevin Newman of Albarrie Naue LTD for their assistance in conducting the tests.

Shear Strength and Long-Term Creep Testing

Robert D. Hewitt[1], Cetin Soydemir[2], Richard P. Stulgis[3], and Matthew T. Coombs[4]

EFFECT OF NORMAL STRESS DURING HYDRATION AND SHEAR ON THE SHEAR STRENGTH OF GCL / TEXTURED GEOMEMBRANE INTERFACES

REFERENCE: Hewitt, R. D., Soydemir, C., Stulgis, R. P., and Coombs, M. T. **"Effect of Normal Stress during Hydration and Shear on the Shear Strength of the GCL/Textured Geomembrane Interfaces,"** *Testing and Acceptance Criteria for Geosynthetics Clay Liners, ASTM STP 1308,* Larry W. Wells, Ed., American Society for Testing and Materials, 1997.

ABSTRACT: A laboratory testing program was performed to evaluate the interface shear strength of a geosynthetic clay liner (GCL) / textured geomembrane interface utilizing two pre-shear inundation methods designed to simulate field conditions. Two commercially available products were tested, a needlepunched and a stitch-bonded GCL. Oedometer swell tests provided swell data for the two products which were used to design the interface shear testing program. Interface shear tests were performed for 1) GCL samples inundated under a low normal stress for a short time and sheared under a higher normal stress, and 2) GCL samples inundated for a longer period under the design normal stress. The results for the two different GCL materials and the two pre-shear inundation conditions are compared.

KEYWORDS: geosynthetic clay liner, GCL, inundation, hydration, interface shear

Liner system design for waste containment facilities warrants investigation into the use of a GCL rather than compacted low-permeability soil, due to its superior hydraulic barrier characteristics under differential settlement and freeze-thaw [1, 2]. As a result, stability of the liner system is often sensitive to the shear strength characteristics of the GCL, both internally and at the interface with the geomembrane. While interface shear characteristics have been published for testing conducted for cover systems at low normal stress levels, little data are available for interfaces tested under the higher normal stress levels

[1] Engineer, Haley & Aldrich, Inc., Cambridge, Massachusetts, 02141.
[2] Vice President, Haley & Aldrich, Inc., Cambridge, Massachusetts, 02141.
[3] Vice President, Haley & Aldrich, Inc., Manchester, New Hampshire, 03310.
[4] Research Assistant, Tufts University, Medford, Massachusetts, 02155.

experienced by a liner. A laboratory testing program was conducted with two commercially available GCLs to establish interface shear resistance under normal stress levels of 69 to 414 kPa simulating a moderate height landfill with up to 30 m of waste fill. Two different pre-shear hydration conditions were investigated in the laboratory program. The first hydration case modeled the field conditions anticipated during construction of a liner to reflect a substantial increase in the water content of the GCL. Hydration of the GCL under low normal stress was followed by a relatively quick loading to simulate placement of waste. The second pre-shear hydration case simulates field conditions where moisture is introduced to the GCL / geomembrane interface at a later stage of waste filling.

MATERIALS

Since the shear resistance mobilized at a GCL / geomembrane interface is product specific, commercially-available geosynthetics were selected for use in the testing program.

Geomembrane

A 40-mil textured Linear Low Density Polyethylene (LLDPE) geomembrane, manufactured by the National Seal Corporation, was used for the interface shear tests. Interface testing was conducted in the machine direction.

Geosynthetic Clay Liners (GCLs)

Two commercial GCLs were selected for laboratory testing: a needlepunched GCL (Bentofix NW), and a stitch-bonded GCL (Claymax 506 SP). Bentofix NW consists of sodium-activated bentonite sandwiched between two needlepunched, non-woven geotextiles. Claymax 506 SP incorporates sodium bentonite mixed with an adhesive between a woven and a non-woven geotextile, stitch-bonded together with polypropylene yarn at four inches on-center. Interface shear tests with Bentofix NW were conducted with the thicker non-woven geotextile at the geomembrane interface. Testing with Claymax 506 SP was conducted with the non-woven geotextile at the geomembrane interface and two rows of stitching along the length of the sample. In the remainder of this paper, Bentofix NW and Claymax 506 SP will be referred to as the needlepunched GCL and the stitch-bonded GCL, respectively.

PROCEDURE

Oedometer swell tests were performed to investigate the swelling characteristics of the two GCLs. Results of the swell test program were used to design the direct shear testing program.

Oedometer Swell Tests

Oedometer swell tests were performed in conventional oedometer testing apparatus in general conformance with ASTM D2435-90, "Standard Test Method for One-Dimensional Consolidation Properties of Soils." The GCL samples were cut from the central portion of the larger roll provided by the manufacturer. The stitch-bonded GCL samples did not include the stitching. The edges of the sample were wetted prior to cutting to minimize loss of bentonite during cutting; samples were weighed after cutting to verify that the GCL samples had not deviated from the specified unit weight. After the samples were placed in the test assembly at their as-delivered water content (see Table 1), the vertical stress was applied and held for 48 hours. At the end of the compression stage, tap water was introduced to the sample, and the vertical displacement (volume change) of the sample was measured until it had approximately reached a stable state. Individual tests were performed for a range of vertical stresses between 6.9 kPa and 345 kPa to provide data on the rate of hydration, thickness, and water content of the GCLs. Water content was determined following ASTM D2216-92, "Standard Test Method for Laboratory Determination of Water (Moisture) Content of Soil and Rock."

Table 1--Summary of oedometer swell tests

GCL	NORMAL STRESS	SAMPLE THICKNESS			DISPLACMENT		WATER CONTENT		
		INITIAL	AFTER COMPRES.	AFTER SWELL	DURING COMPRES.	DURING SWELL	INITIAL	AFTER SWELL	CHANGE
	kPa	mm	mm	mm	mm	mm	%	%	%
NP	6.9	7.247	7.018	9.753	0.229	-2.735	11.8	176.1	164.2
NP	34.5	7.772	7.533	9.943	0.239	2.410	9.4	154.1	144.7
NP	69.0	7.841	7.477	9.210	0.364	-1.733	9.0	115.5	106.5
NP	137.9	7.882	7.494	8.557	0.388	-1.063	11.3	119.9	108.6
NP	206.9	7.816	7.483	7.902	0.330	-0.419	10.3	110.3	100.1
NP	344.8	7.696	7.221	7.328	0.475	-0.107	15.2	95.0	79.7
SB	6.9	6.561	6.440	12.220	0.121	-5.780	7.9	246.5	238.6
SB	34.5	6.934	6.587	10.907	0.347	-4.320	15.7	207.7	192.0
SB	69.0	6.942	6.582	9.733	0.360	-3.151	15.2	171.8	156.6
SB	137.9	6.139	5.856	7.713	0.283	-1.857	6.7	130.0	123.3
SB	206.9	6.147	5.854	6.979	0.293	-1.125	6.8	115.4	108.6
SB	344.8	6.104	5.867	6.279	0.237	-0.412	7.0	91.5	84.5

Interface Shear Tests

Interface shear testing was conducted in general accordance with ASTM D5321-92, "Standard Test Method for Determining the Coefficient of Soil and Geosynthetic or Geosynthetic and Geosynthetic friction by the Direct Shear Method." Large scale direct shear tests were performed in a Boart-Longyear LG-101 Large Direct Shear Machine which utilizes a bladder system to apply normal stress. The bladder system was calibrated during the testing program by comparing the applied air pressure and the normal load measured at the interface with load cells. End clamps and textured aluminum plates were used to secure the GCL and geomembrane in the shear box so that slippage would not occur between the geosynthetics and the shear box. All interface testing was performed with the non-woven geotextile in contact with the geomembrane. Samples were examined during and after shear to inspect for visible distortion of the GCL or geomembrane samples. Water content and thickness of small coupons cut from the larger sample were determined at various stages of the testing (Table 2). Water content was measured

Table 2--Summary of interface shear tests

GCL	TIME OF INUNDATION	NORMAL STRESS		SHEAR STRESS		NORMALIZED SHEAR RESISTANCE		WATER CONTENT				THICKNESS	
		DURING INUNDATION	DURING SHEAR	PEAK	AT 75 mm DISPL.	PEAK	AT 75 mm DISPL.	INITIAL	FINAL AT CENTER	FINAL AT SIDE	FINAL AVERAGE	INTIAL	FINAL
	DAYS	kPa	kPa	kPa	kPa			%	%	%	%	mm	mm
NP	2	6.9	69.0	41.4	26.0	0.60	0.38	5.1	140.0	---	---	---	---
NP	2	6.9	206.9	103.4	35.0	0.50	0.17	5.1	104.0	---	---	---	---
NP	2	6.9	310.3	131.0	36.0	0.42	0.12	---	---	---	---	---	---
NP	15	103.4	103.4	66.0	41.4	0.64	0.40	5.7	91.0	113.7	102.4	7.8	8.6
NP	15	206.9	206.9	100.0	65.5	0.48	0.32	4.5	72.4	95.3	83.9	7.8	8.1
NP	15	413.7	413.7	200.0	117.2	0.48	0.28	5.9	73.6	84.4	79.0	8.1	8.2
SB	3	6.9	69.0	41.0	30.0	0.59	0.44	29.0	250.0	---	---	---	---
SB	2	6.9	206.9	61.0	55.0	0.29	0.27	7.1	161.1	215.0	188.1	6.8	10.4
SB	2	6.9	344.8	71.0	57.0	0.21	0.17	7.1	150.6	271.6	211.1	6.8	10.4
SB	12	103.4	103.4	54.5	40.0	0.53	0.39	10.7	96.8	134.0	115.4	6.9	9.1
SB	10	206.9	206.9	80.0	65.0	0.39	0.31	7.1	76.4	119.3	97.8	6.8	8.1
SB	6	413.7	413.7	108.0	70.0	0.26	0.17	7.1	60.4	91.2	75.8	6.9	7.2

using ASTM D2216-92 procedures and the thickness of the GCL was determined with a micrometer. Shearing was conducted at a rate of 1 mm/min and continued to a displacement of at least 76 mm. Samples were inundated under two different procedures to represent the range of potential field conditions that could be experienced by the GCL. The two procedures are described below.

Case 1—To simulate inundation of the GCL under a low normal stress level during construction, prior to placement of waste, GCL samples were inundated outside of the direct shear machine for two days under a normal stress of 6.9 kPa. The normal stress was applied with dead weights. Geonet was placed on both sides of the GCL during inundation to promote uniform hydration similar to the oedometer test setup. After the soaking period, the samples were carefully transferred to the shear box to prevent contaminating the interface with bentonite. The normal stress, ranging from 69 kPa to 345 kPa, was then applied in increments over 1 to 2 hours. Shearing was initiated approximately 15 minutes after the last normal stress increment was applied to approximate undrained shear. During shear the samples had no access to water.

Case 2—To model a GCL maintained at the manufactured moisture during construction, but then inundated at some time after application of a large overburden stress, GCL samples were placed in the shear box at their as-delivered water content and the full normal stress of the test was applied. The normal stress was incrementally applied over a period of 1 to 2 hours, prior to flooding the shear box with tap water to a level approximately 13 mm above the top of the GCL. This water level was maintained during the 8 to 16 day soaking period. The length of the soaking period was selected, based on the results of the oedometer swell tests, to allow the GCL to attain at least 90% of its total swell displacement prior to shear. At the end of the soaking period the shear box was drained of free water and shearing was initiated with no access to free water.

LABORATORY TEST RESULTS

Oedometer Swell Tests

Results of the oedometer swell tests are presented in Figures 1 through 5 and summarized on Table 1. The displacement of the GCL (volume change) is plotted versus the time after the addition of water in Figure 1 for the needlepunched GCL and Figure 2 for the stitch-bonded GCL. The amount of compressive displacement during the loading period prior to the addition of water is indicated on Table 1. It was observed that for both GCLs the swell tests performed at a normal stress of 344.8 kPa continued to compress upon the addition of water before eventually swelling. Summary plots of the vertical strain (ε_v) versus normal stress level for both GCLs are presented in Figure 3.

$$\varepsilon_v\ (\%) = (\text{Total Swell Displacement} / \text{Thickness after Compression}) * 100$$

The final water content at the conclusion of the test was plotted versus normal stress for the needlepunched and the stitch-bonded GCLs in Figures 4 and 5, respectively.

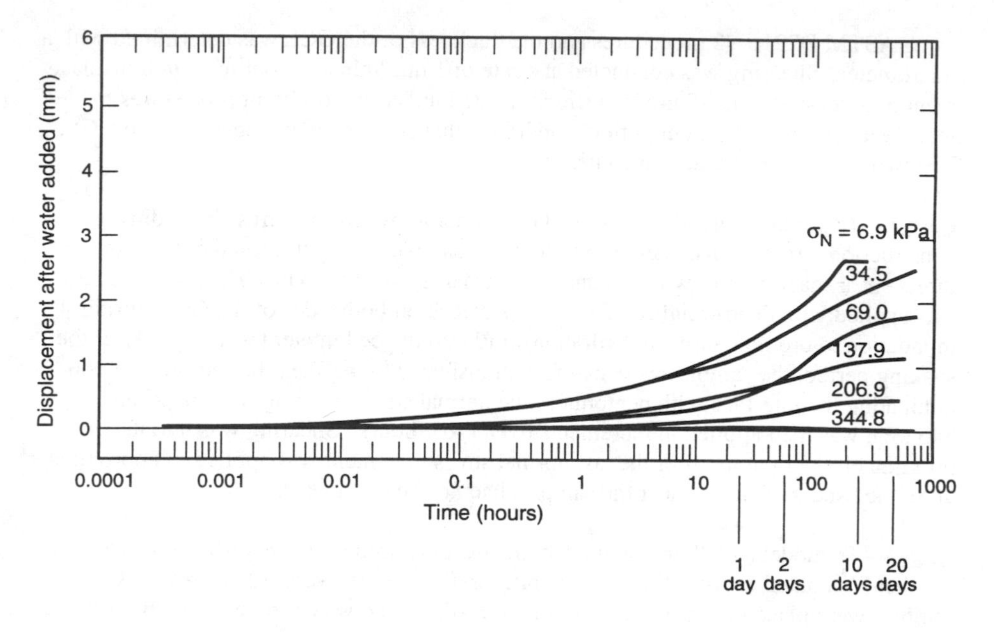

FIG. 1--Swell tests of needlepunched GCL specimens

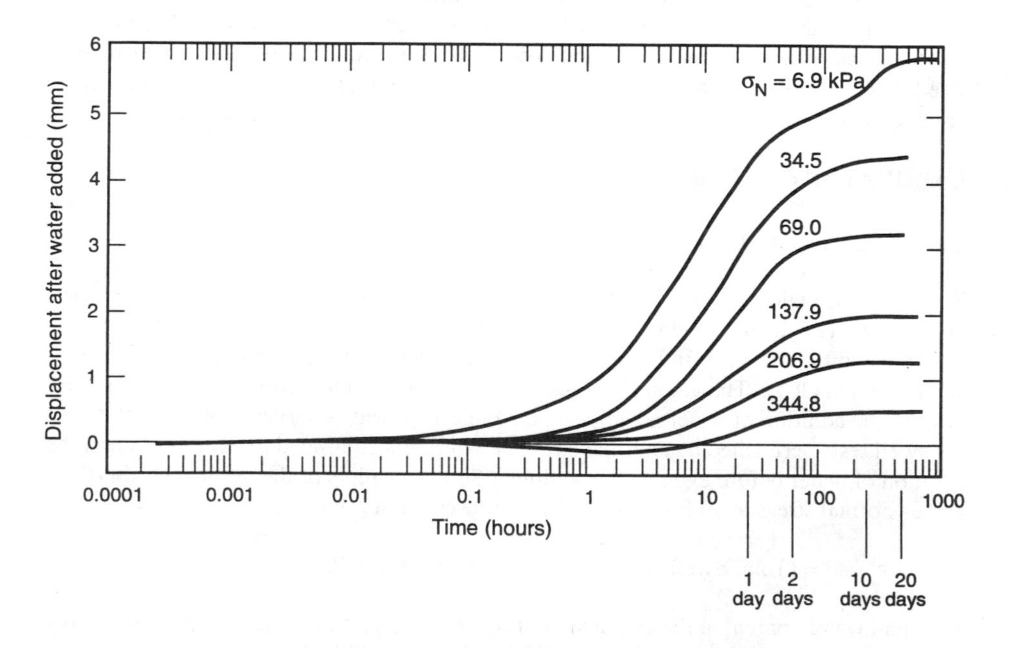

FIG. 2--Swell tests of stitch-bonded GCL specimens

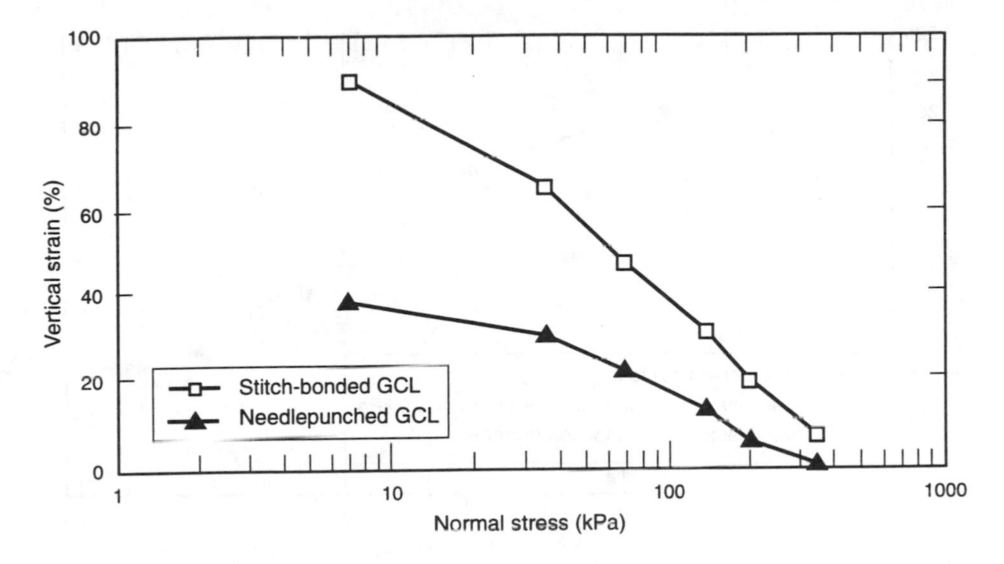

FIG. 3--Vertical strain in oedometer swell tests after the addition of water versus normal stress

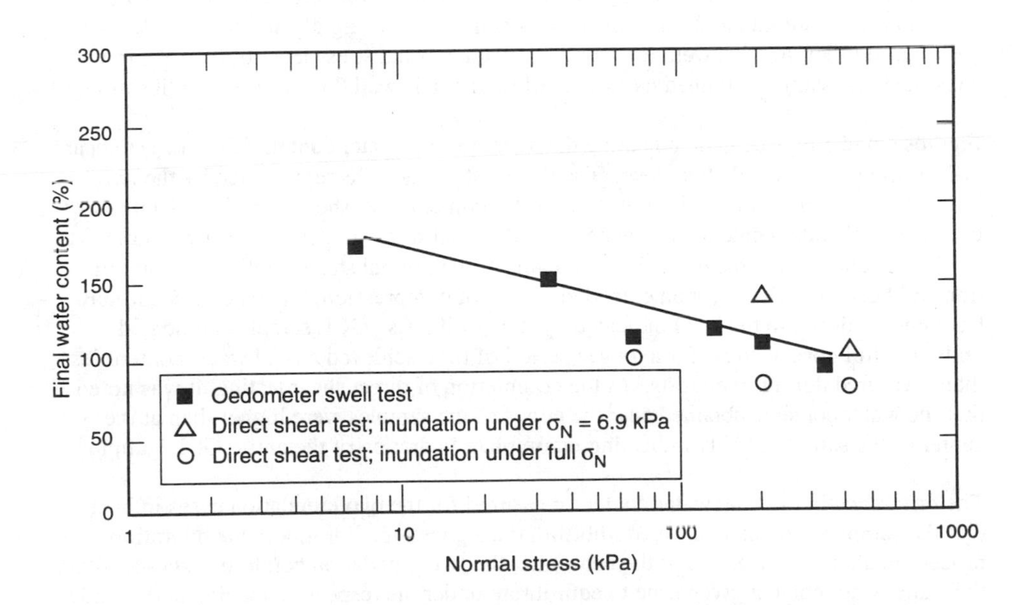

FIG. 4--Final water content versus normal stress for needlepunched GCL

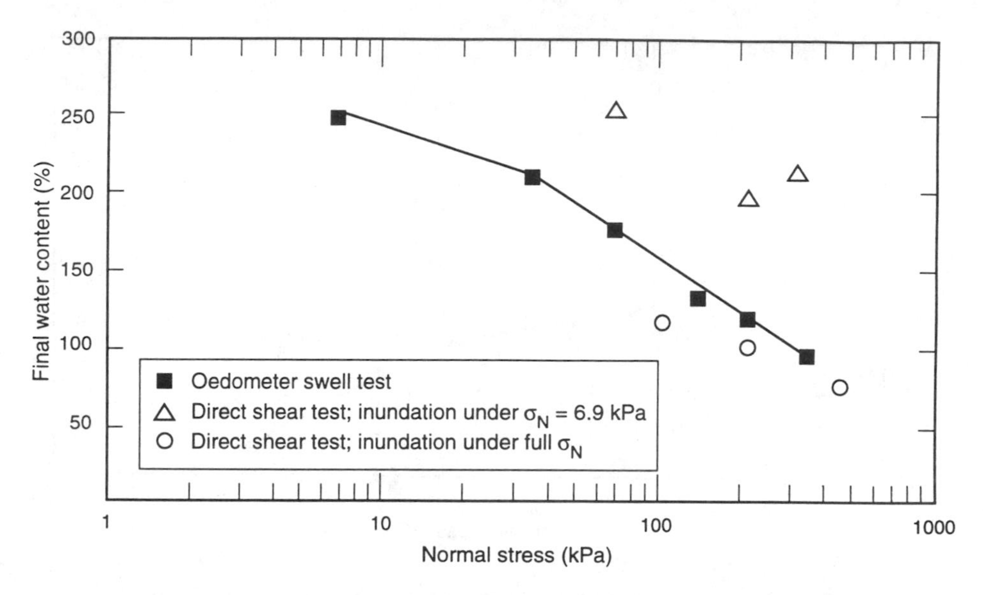

FIG. 5--Final water content versus normal stress for stitch-bonded GCL

The results of the oedometer tests indicate that for the tested range of normal stresses, the GCLs achieved at least 90% of their total swell displacement (volume change) in about 10 days. In general, as the normal stress increased, the total swell displacement decreased, a related reduction in water content of the sample was observed, and a stable condition (little additional displacement) was achieved in a slightly shorter time of soaking. Further, for the swell tests conducted at a normal stress level of 6.9 kPa, in two days the GCL samples attained 64 to 76% of their final swell thickness at equilibrium.

Figures 4 and 5 include data indicating the average final water content of the direct shear GCL samples, measured after shear. The different stress paths represented by the two inundation cases result in a different final water content after shear. For Case 1 inundation, the final water content was greater than the final water contents measured in the swell tests under the same normal stress level. If the normal stress applied after inundation had been maintained for an extended period of compression, the water content may have been reduced to the level indicated by the swell tests. GCL samples inundated under the full normal stress for a longer period of time achieved a final water content less than measured during swell tests. At the completion of direct shear testing, it was noted that the water contents obtained near the edges of the samples were higher than at the center of the samples, likely indicating incomplete hydration of the entire GCL sample.

The difference in the final water content measured for the two inundation cases indicates that the samples did not achieve equilibrium during testing. For unless the hydration procedure alters the structure of the bentonite, the two inundation conditions should attain the same water content given time to equilibrate under the respective loading and inundation conditions.

Interface Shear Tests

Interface shear test results are presented on Figures 6 through 13 and in Table 2. Test results for the needlepunched GCL and the stitch-bonded GCL are discussed separately, followed by a comparison of the test results for the two GCLs.

Needlepunched GCL—For the needlepunched GCL / geomembrane interface inundated under the full normal stress (Case 2), the strain-softening behavior of the interface becomes more pronounced with increasing normal stress. As the normal stress increases the initial slope of the shear stress versus horizontal displacement curve, the peak shear stress, and the large displacement shear stress increases (Figure 6).

The same interface tested with pre-shear inundation under low normal stress followed by application of a high normal stress (Case 1) exhibited less defined trends. The peak shear stress and initial slope of the shear stress versus horizontal displacement curve (Figure 7) increase with normal stress. However, large displacement shear stress values were similar for the two tests conducted at the highest normal stress levels. Further, all three tests exhibited different shear stress versus displacement curves. It was noted that during application of the highest normal stress, bentonite and water were forced from the sides of the test sample, potentially changing the nature of the interface.

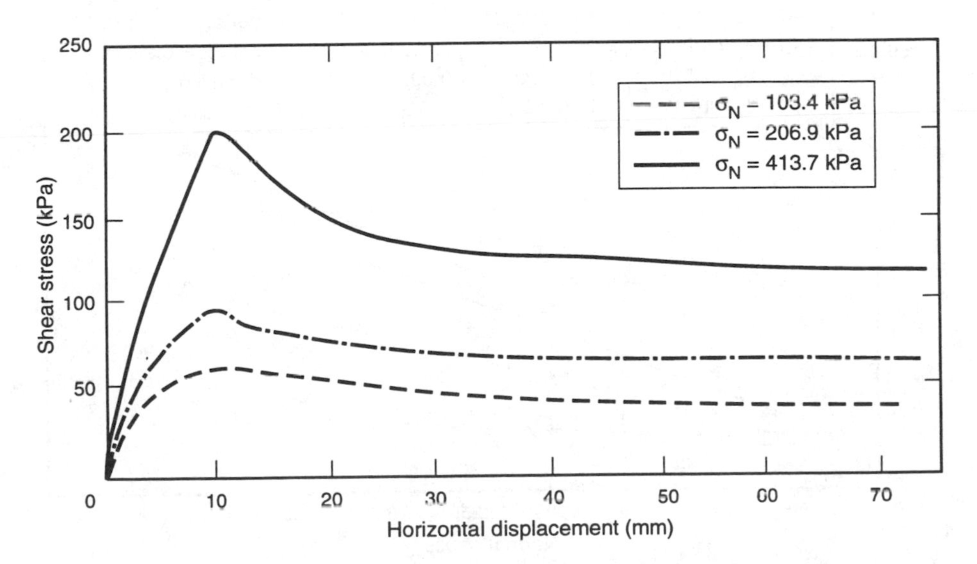

FIG 6.--Shear stress versus horizontal displacement for needlepunched GCL / textured geomembrane interface inundated under the full normal stress

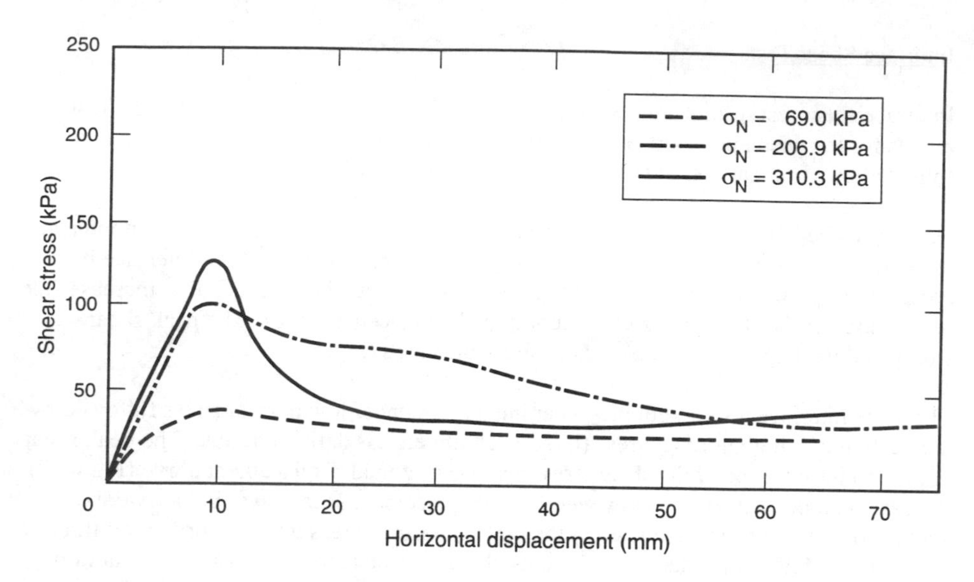

FIG. 7--Shear stress versus horizontal displacement for needlepunched GCL / textured geomembrane inundated under 6.9 kPa

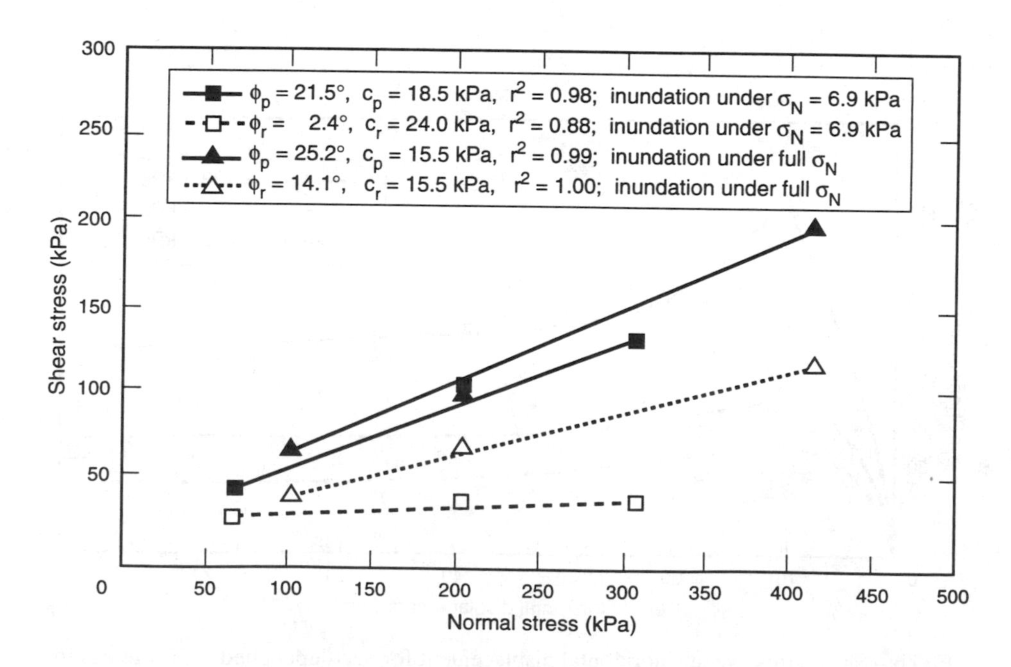

FIG. 8--Failure envelopes for needlepunched GCL / textured geomembrane interface

A comparison of the needlepunched GCL / geomembrane interface tested under the two methods of inundation is illustrated in Figure 8, which presents failure envelopes for the peak and large displacement shear resistance. Linear regression parameter (c, ϕ , r^2) are summarized on the figure, where r^2 indicates the relative quality of the line fit. While the interface inundated under the full normal stress (Case 2) achieved marginally higher peak shear resistance than Case 1, the increase in large displacement shear resistance inundation was substantial. In general, the normalized shear resistance decreased with increasing normal stress (Figure 9). For both inundation cases, peak shear stress values were mobilized at a displacement of 8 to 10 mm.

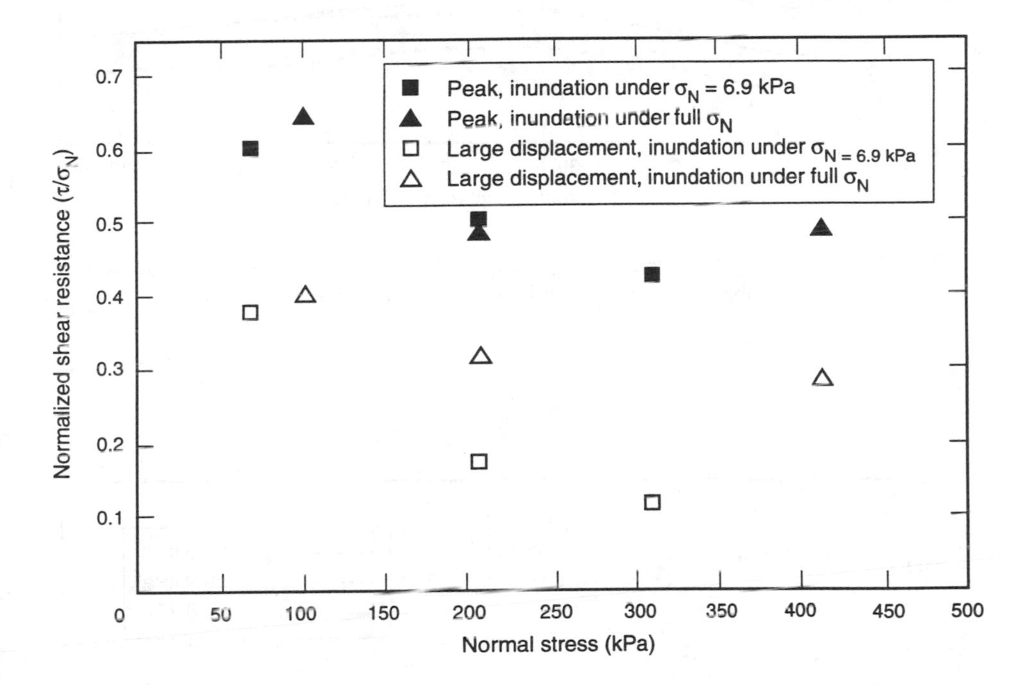

FIG. 9--Normalized shear resistance (τ/σ) versus normal stress for needlepunched GCL / textured geomembrane interface

Stitch-bonded GCL—The stitch-bonded GCL / textured geomembrane interface submerged in water under the full normal stress (Case 2) achieved an initial peak shear stress followed by a slight decrease. The shear stress then increased gradually to a higher peak value before decreasing to a steady large displacement shear stress. The highest peak value was attained at a displacement of 28 to 38 mm. The initial slope of the shear stress versus horizontal displacement curve (Figure 10) and the peak shear stress increased with applied normal stress; however, the large displacement shear stress for the two sample points at the highest normal stress levels were similar despite a difference in normal stress level.

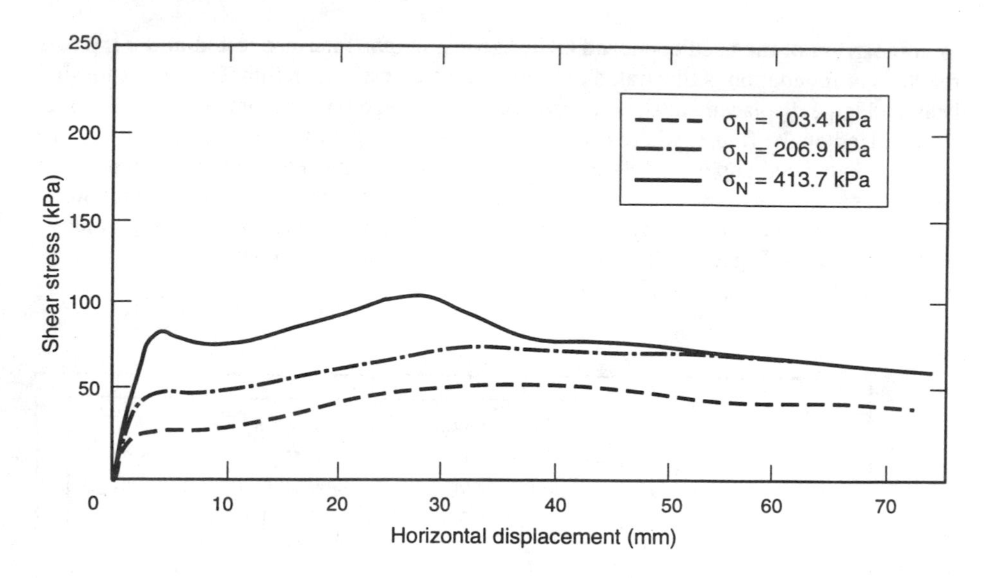

FIG. 10--Shear stress versus horizontal displacement for stitch-bonded GCL / textured geomembrane interface inundated under the full normal stress

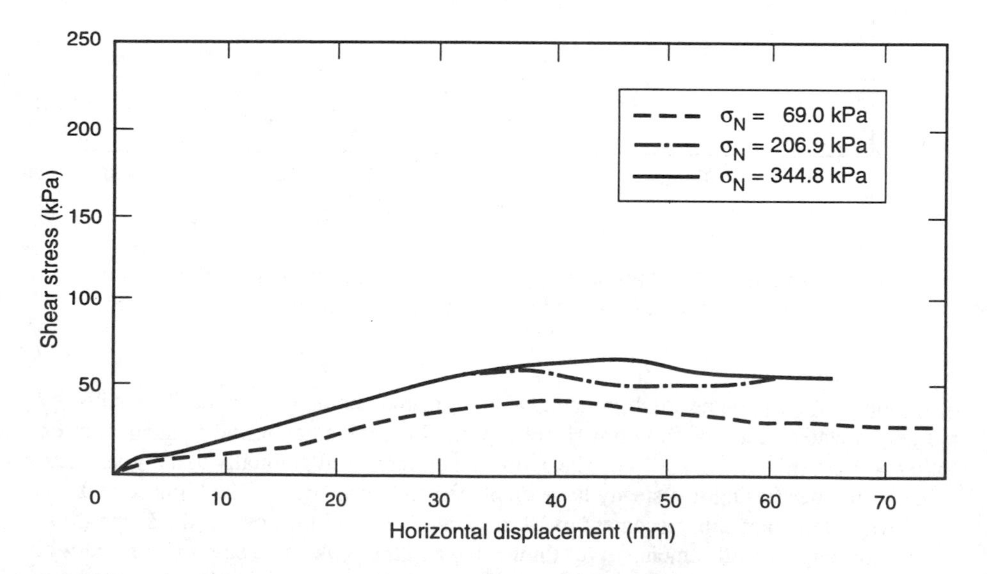

FIG. 11--Shear stress versus horizontal displacement for stitch-bonded GCL / textured geomembrane interface inundated under low normal stress (6.9 kPa)

For the same interface flooded under inundation Case 1, the shear stress gradually increased to a peak value followed by a gradual decrease in shear resistance (Figure 11). No initial peak behavior as noted above for Case 2 was observed. The peak shear stress was mobilized at a horizontal displacement of 35 to 48 mm. Peak shear stress values increased with normal stress. However, similar to Case 2, the shear resistance at large displacement for the two tests conducted at the highest normal stress levels were similar. Strength envelopes for the large displacement shear resistance were similar for the two inundation methods (Figure 12). The tests conducted with samples inundated under the full normal stress (Case 2), however, achieved a higher peak strength envelope than the

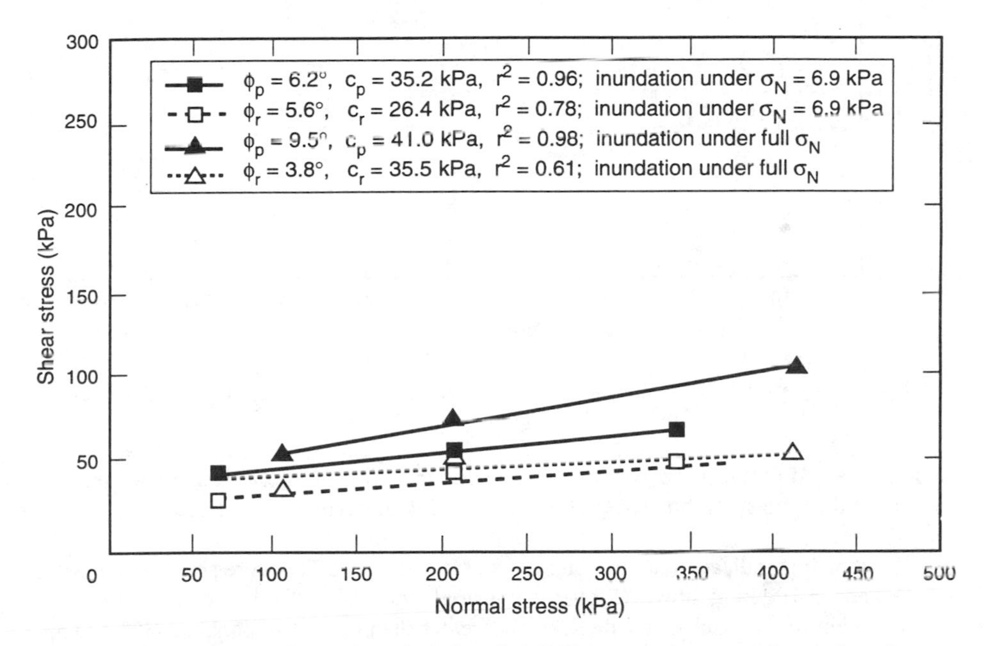

FIG. 12--Failure envelopes for stitch-bonded GCL/textured geomembrane interface

samples inundated under 6.9 kPa normal stress (Case 1). The peak strength envelope for Case 1 was similar to the large displacement strength envelopes. The low r^2 regression values for the large displacement failure envelopes may indicate that a non-linear model is appropriate over a larger range of normal stress, especially for tests conducted below 69 kPa [3].

<u>Comparison of Needlepunched GCL and Stitch-bonded GCL</u>--Mobilization of the shear resistance at the GCL / geomembrane interface exhibits basic differences for the two GCLs tested. Peak shear resistance is mobilized at markedly different displacements and distinctly different magnitudes. Peak shear resistance was mobilized at a displacement of 8 to 10 mm for the needlepunched GCL and from 28 to 48 mm for the stitch-bonded GCL. The normalized shear resistance (τ/σ) decreases with increasing normal stress in the same general trend for the needlepunched and the stitch-bonded GCL, as highlighted

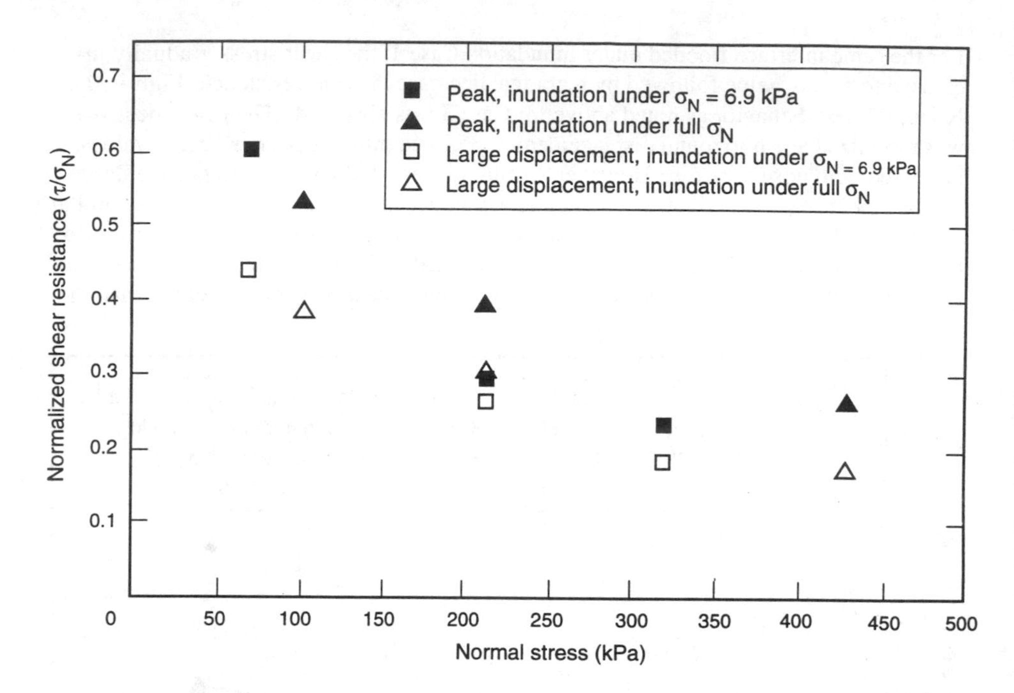

FIG. 13--Normalized shear resistance (τ/σ) versus normal stress for stitch-bonded GCL / textured geomembrane interface

on Figure 9 and Figure 13. However, the needlepunched GCL mobilized greater shear resistance than the stitch-bonded GCL over the range of normal stress tested.

Swell test results indicate that the water content of the two GCLs were similar for normal stresses above 100 kPa; however, over the normal stress range 6.9 kPa to 100 kPa the water content of the stitch-bonded GCL was greater than the needlepunched GCL. The thickness of the GCLs exhibited a similar trend, with the vertical strain induced during swell ranging from 90% at a normal stress of 6.9 kPa to 6% at 344.8 kPa for the stitch-bonded GCL; the needlepunched GCL developed a vertical strain of 40% to 2% over the same normal stress range.

To evaluate the influence of the water content, the final water content of the GCL is compared to the water content at the equilibrium level determined by the swell tests data presented on Figure 4. The samples inundated under the full normal stress had an average water content slightly (10 to 20% water content) less than the swell test data. The interface shear test as performed does not allow for determination of the water content at the conclusion of inundation. Therefore, the discrepancy between the swell tests and the interface shear tests was inferred to be due to the shorter time of swell for the shear box tests (incomplete hydration), longer drainage paths, or partial drainage during shear. However, the trend of the data matches the swell test results.

The water content of stitch-bonded interface shear samples inundated at 6.9 kPa were up to 100% higher in water content than the swell. Likewise, the needlepunched GCL sample inundated under a normal stress of 6.9 kPa were observed to have a final water content in excess of the swell test data, although not as much as the stitch-bonded GCL. The effects of the increased water in the system may include lubrication of the interface during shear, resulting in a reduced shear resistance at the interface.

Further, during the transfer of the samples to the shear machine following inundation, the interface may have been contaminated by fugitive bentonite. However, the similarity of the peak values for both inundation cases with the needlepunched GCL suggest that bentonite contamination due to the sample transfer procedure did not occur for this material.

SUMMARY AND CONCLUSIONS

In summary, laboratory procedures intended to develop the shear resistance of a GCL / geomembrane interface should be modeled to reflect the anticipated field hydration and shearing conditions. Hence, laboratory testing should be an integral part of the design process for critical interfaces within the liner system.

The following conclusions were derived from the results of the laboratory testing program on the GCL / textured geomembrane interface:

1. Inundating the GCLs under a low normal stress (6.9 kPa), followed by application of a higher normal stress (Case 1) during shear produced a lower mobilized shear resistance than tests on samples inundated under the full normal stress of the test (Case 2). Possible explanations for the lower shear resistance at the interface include: 1) Distortion of the sample during application of the normal stress, 2) a higher sample water content, and 3) lubrication of the interface by bentonite lost during sample transfer to the shear machine. Further, the tests inundated under the full normal stress (Case 2) exhibited more consistent trends.

2. Under both inundation cases the test data consistently indicated that the GCL / geomembrane interface exhibited a substantial decrease in strength after peak values of shear stress were attained. The needlepunched GCL exhibited a higher peak and large displacement shear resistance than the stitch-bonded GCL as well as a larger reduction in shear resistance after peak. Since for the needlepunched GCL less distortion of the GCL and less free bentonite was observed at the interface during shear, the densely reinforced structure resulting from the needlepunching process may have contributed to the higher shear resistance of the needlepunched GCL.

3. The needlepunched GCL and the stitch-bonded GCL / textured geomembrane interface attained peak values at widely different horizontal displacements. The interface with the needlepunched GCL mobilized peak shear resistance at a horizontal displacement of 8 to 10 mm, while the interface with the stitch-bonded GCL approached peak shear resistance at horizontal displacements ranging from 25 to 48 mm. The densely reinforced needlepunching may have resulted in a stiffer GCL structure, which exhibited brittle

failure behavior at the GCL / geomembrane interface. The stitch-bonded GCL, with a lower density of reinforcement, behaved as a more ductile material at the interface with the geomembrane.

4. Both the needlepunched GCL and the stitch-bonded GCL interfaces follow the trend of decreasing normalized strength (τ/σ) with increasing normal stress. Modeling of the failure envelopes, especially for large displacements, may require non-linear failure envelopes to characterize the variation of shear resistance over the full range of design normal stress [3].

The interface shear strength parameters indicated on Figures 8 and 12 are only valid for the laboratory inundation and shearing conditions of the tests reported in this paper. For design, interface shear tests should be conducted under the anticipated field conditions.

While the internal strength of GCLs was not considered in this investigation, the critical failure surface may shift from the interface to the midplane of the GCL as the normal stress varies throughout a design section [4].

The shear rate of testing was not varied during the laboratory testing program; however, the shear rate of testing may influence the mobilized shear resistance [5].

ACKNOWLEDGEMENTS

The work of Ms. Acey Welch during the preparation of this paper along with the assistance of Mr. Mark Dobday, Mr. Desmond Crawford, and Ms. Freddie Farrar during the laboratory testing program is greatly appreciated by the authors.

REFERENCES

[1] Stulgis, R. P., Soydemir, C., Telgener, R. J., and Hewitt, R. D., "Use of Geosynthetics in 'Piggyback Landfills'": A Case Study," Procedings of GRI Conference, January 1995.

[2] Daniel, D. E. "Soil Barrier Layers Versus Geosynthetic Barriers in Landfill Cover Systems," in Proc. Landfill Closures...Environmental Protection and Land Recovery, Ed. R. Jeffrey Dunn and Singh, Udai P., Geotechnical Special Publication No. 53, 23 - 27 October 1995.

[3] Stark, Timothy D. and Poeppel, Alan R. "Landfill Liner Interface Strengths from Torsional-Ring-Shear Tests," ASCE Journal of Geotechnical Engineering, vol. 120, no. 3, March 1994.

[4] Gilbert, R. B., Fernandez, F., and Horsfield, D. "Shear Strength of a Reinforced Geosynthetic Clay Liner," submitted to the ASCE Journal of Geotechnical Engineering.

[5] Daniel, D.E., Shan, H.-Y., and Anderson, J.D. (1993), "Effects of Partial Wetting on the Performance of the Bentonite Component of a Geosynthetic Clay Liner," Proceedings of Geosynthetics '93, Vancouver, B. C.,

Kelly S. Merrill [1] and Arthur J. O'Brien [2]

STRENGTH AND CONFORMANCE TESTING OF A GCL USED IN A SOLID WASTE LANDFILL LINING SYSTEM

REFERENCE: Merrill, K. S. and O'Brien, A. J., **"Strength and Conformance Testing of a GCL Used in a Solid Waste Landfill Lining System,"** *Testing and Acceptance Criteria for Geosynthetic Clay Liners, ASTM STP 1308,* Larry W. Well, Ed., American Society for Testing and Materials, 1997.

ABSTRACT: This paper describes strength and conformance tests conducted on a Bentomat ST geosynthetic clay liner (GCL) used in a composite lining system for the Cells 4 and 5 expansion of the Anchorage Regional Landfill in Anchorage, Alaska. The Cells 4 and 5 lining system included use of an 80-mil, high-density polyethylene (HDPE) liner overlying a GCL on both the sideslopes and base of the cells. The use of this lining system in a Seismic Zone 4 area on relatively steep side slopes required careful evaluation of both internal shear strength of the GCL and interface friction between the GCL and textured HDPE. Laboratory tests were carried out to evaluate both peak and residual GCL internal strengths at normal loads up to 552 kiloPascals (80 pounds per square inch). Laboratory tests also were conducted to evaluate the interface strength between the GCL and Serrot "box and point" textured HDPE. Interface strengths between both woven and nonwoven sides of the GCL and the textured HDPE were evaluated. Considerations related to use of peak or residual strengths for various interim stability cases are described in this paper. Stability analyses using stress-dependent interface and internal strengths for the GCL are addressed. The quality assurance and conformance testing program adopted for the project on GCL is discussed also.

KEYWORDS: GCL, interface strengths, internal shear strength, solid waste landfill, composite lining system, textured HDPE, conformance tests, slope stability, seismicity

---

[1] Geotechnical Engineer, CH2M HILL, 301 W. Northern Lights Blvd., Suite 501, Anchorage, AK 99503

[2] Environmental Business Group Manager, CH2M HILL, 2485 Natomas Park Drive, Suite 600, Sacramento, CA 95833

1.0 BACKGROUND

The Anchorage Regional Landfill (ARL) is a solid waste municipal landfill located about 10 miles north of Anchorage, Alaska. The ARL opened in 1987 with Cell 1, comprising approximately 25 acres. Cell 2 (15 acres) and Cell 3 (10 acres) were brought into operation in 1990 and 1992, respectively. This paper describes strength and conformance testing for a composite high-density polyethylene (HDPE)/geosynthetic clay liner (GCL) lining system installed as part of the most recent landfill expansion, Cells 4 and 5 (23 acres), which was completed in fall 1995 and is now receiving waste. The Cells 4 and 5 design included use of an HDPE/GCL lining system on both the side slopes and the base of the cells due to the lack of a reliable clay source in the Anchorage area. Considerations related to use of GCLs in similar landfill applications are described in References [1] through [7]. To assess the stability of the lining system in accordance with U.S. Environmental Protection Agency (EPA) Subtitle D requirements in a Seismic Zone 4 area on relatively steep side slopes (2.7 horizontal to 1 vertical), project-specific tests were carried out to determine the HDPE/GCL interface strength and the GCL internal shear strength. This paper documents the testing program carried out to obtain realistic strength properties for stability assessments, briefly describes the stability evaluations, and discusses the conformance testing program implemented to confirm acceptable material properties.

2.0 STRENGTH TESTING PROGRAM

The strength testing program included determination of HDPE/GCL interface strengths and GCL internal shear strength. The tests were carried out on the geosynthetic materials proposed for use on the project by the geosynthetics subcontractor. The primary objectives of the testing program were as follows:

1. To confirm design assumptions made concerning geosynthetic strength parameters
2. To aid in determination of the maximum allowable height of refuse from a slope stability standpoint
3. To aid in evaluation of placement orientation of the GCL

The HDPE proposed for the project was 80-mil, single-sided, textured material from Serrot Corporation. The Serrot HDPE texturing consists of the "box and point" grid system, with raised intersecting ribs forming a regular pattern of square grids approximately 1.3 cm (1/2 inch) across. A single raised point is located in each square grid. The texturing is formed integrally with the parent material during the extrusion process; the reported advantage is more consistent texturing than in materials where the texturing is added after the parent material is manufactured. The HDPE was specified as single-sided, with the textured side placed down and the smooth side up. The intent of this orientation is to provide a predetermined slip plane above the lining system on the side slopes, where a woven reinforcing geotextile was to be placed directly above the HDPE to support the overlying sand cover material.

The GCL proposed for use on the project was Bentomat ST from Colloid Environmental Technologies Company (CETCO), which consists of sodium bentonite clay sandwiched between a woven polypropylene geotextile on one side and a nonwoven polypropylene geotextile on the other. During manufacture of this material, needle-punching equipment pushes fibers from the nonwoven geotextile through the bentonite clay and into the woven geotextile to provide internal reinforcement and bentonite encapsulation. Descriptions of test program methodology and results are presented in the following paragraphs.

2.1 Interface Friction Tests

Laboratory direct shear tests were conducted to determine the interface shear strength between the Bentomat ST GCL and the textured side of the single-sided Serrot 80-mil HDPE. Because of time constraints, the tests were conducted on samples of the geosynthetics certified to be similar to the material actually used on the project, not on samples from rolls delivered to the site. All tests were conducted by GeoSyntec Consultants in Atlanta [8]. The tests were conducted in general accordance with American Society for Testing and Materials (ASTM) D 5321 [9], although specific test parameters were assigned to best model anticipated field conditions as follows:

- Each test was run under hydrated and consolidated conditions. It is known that GCLs tend to hydrate relatively quickly, even on dry subgrades [2]. Additionally, it is anticipated that consolidation of the thin GCL will occur relatively rapidly as refuse load is applied. Accordingly, it appears reasonable to test the sample under hydrated and consolidated conditions to most accurately model field conditions.
- Each specimen was hydrated with tap water, including both GCL and HDPE components together, for 24 hours under a normal stress of 69 kiloPascals (kPa) (10 pounds per square inch [psi]). This procedure is intended to model expected GCL hydration occurring under a load less than the loads under which the material will be sheared. Initial moisture contents for the GCL samples ranged from 20 percent to 23 percent; final moisture contents ranged from about 130 percent to 200 percent, indicating a fairly high degree of hydration was achieved when compared to similar values achieved in hydration and swell tests conducted by Daniel [1].
- After hydration, each sample was consolidated at the designated normal stress for 12 hours prior to being sheared. For the first specimen tested, the deformation versus time during consolidation was monitored to confirm that a 12-hour period was sufficient to achieve primary consolidation. Based on the results of this initial test, primary consolidation was effectively achieved within about 10 hours (see Figure 1).
- Each interface was sheared at three normal stresses (138, 276, and 552 kPa [20, 40, and 80 psi]). These normal stresses are equivalent to approximately 18 to 76 meters (60 to 250 feet) of refuse, or the likely stress range for the critical stability cases being analyzed. Shearing was accomplished by transferring load to the GCL/HDPE interface from an upper shear box. A textured steel gripping surface was placed on top of the GCL to minimize slippage between the GCL and the rigid wooden substrate of the upper shear box. The GCL was attached to the upper shear box with mechanical clamps. The HDPE was attached to an immobile lower shear box with mechanical clamps.

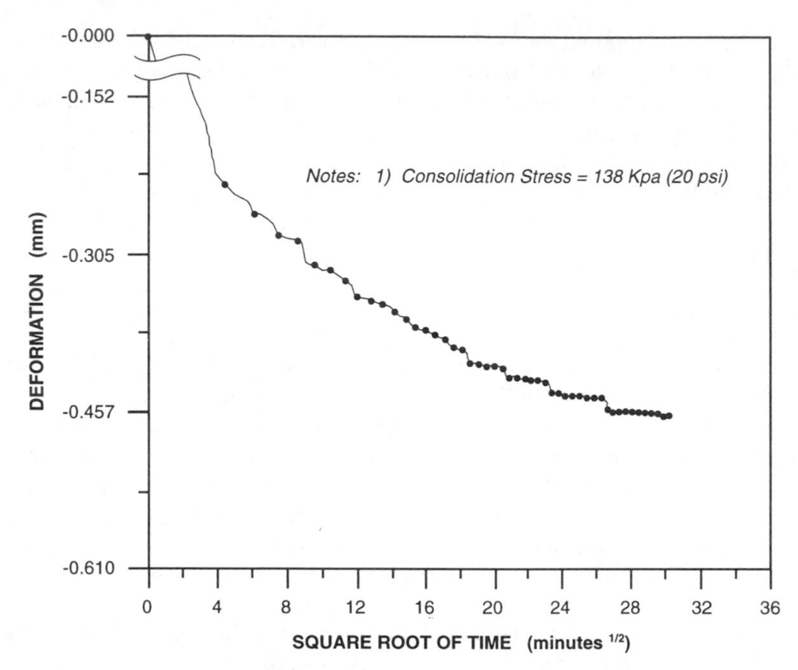

FIG. 1 -- Deformation vs. Time - Consolidation Stage
Textured HDPE/GCL (Woven Side)
Interface Direct Shear Test

- Peak and residual strength values were obtained at each normal stress for each specimen. The residual values were determined from shear force—displacement curves as the post-peak point where the curve stabilizes to a near constant force with increasing displacement. The residual strength values obtained are reported with the associated horizontal displacement values, because continued displacement beyond the available range of the direct shear equipment used here (approximately 8 cm [3 inches]) could potentially produce slightly lower residual values. Accordingly, the reported residual strength values should always be considered in conjunction with the associated displacement values.
- Specimen size was 30 cm by 30 cm (12 inches by 12 inches).
- Shear rate was a constant 0.10 cm (0.04 inch)/minute. This value was chosen to be consistent with shear rates typically used for other available interfacial strength data for comparative purposes and to achieve specimen failure within a reasonable and cost-effective time period (approximately 1 hour or less). There is little information concerning the effects of shear rate on interface strengths, although the relatively rapid shear rate used for these tests is not expected to result in fully drained conditions in cases where the bentonite from the GCL extrudes through the geotextile into the interface. Shear rate impacts on internal GCL strengths are discussed in a subsequent section.

2.1.1 Textured HDPE/GCL Woven Side--One set of interface friction tests was run with the woven side of the Bentomat ST in contact with the textured HDPE. This is the standard installation orientation of the Bentomat ST geotextile, with the woven side up and the nonwoven side down to maximize the interface friction with the underlying subgrade. The direction of shear for all tests was in the machine direction of the geosynthetics. Peak and residual strengths for the HDPE/GCL woven side interface are presented graphically in Figure 2. The results are summarized in Table 1, along with available existing data concerning the interface friction between the textured HDPE and the woven side of the Bentomat ST GCL. The data from this project are generally within the range of the previous data, although the existing peak strength data available for comparison are limited.

2.1.2 Textured HDPE/GCL Nonwoven Side--A second set of tests was run with the nonwoven side of the Bentomat ST in contact with the textured HDPE. These tests were run to determine whether the nonwoven side will provide a higher interface shear strength due to less extrusion of bentonite through the nonwoven fabric and greater frictional interaction between the nonwoven and the textured HDPE. If so, consideration would have been given to installing the GCL with the nonwoven side up to take advantage of the increased interface strength, thus enhancing the sliding stability of the lining system. This decision would also depend on an assessment of the increased lateral transmissivity resulting from placing the nonwoven geotextile immediately below the HDPE liner instead of the woven geotextile.

The test results for the textured HDPE/GCL nonwoven side are shown graphically in Figure 3. Both peak and residual strengths for the textured HDPE/GCL nonwoven interface are also included in Table 1. The results indicate that for the samples tested in this study, the HDPE/GCL nonwoven interface has a slightly higher peak friction than does the HDPE/GCL woven interface. However, the HDPE/GCL woven interface has a slightly higher residual strength up to a normal stress of about 517 kPa (75 psi), or through most of the stress range considered for this study. Because of the similarity of results for the two interfaces tested, the GCL was installed woven side up in accordance with the manufacturer's recommended standard orientation.

It should be noted that the results of the interface tests comparing the woven and nonwoven sides of the GCL were most likely influenced by the "box and point" HDPE texturing, and other types of HDPE texturing may provide different relative results. It is theorized that the nonwoven interface provides a higher peak strength because of greater penetration of the "box and point" textured grid at low deformations. At the higher deformations of the residual case, however, the greater interlocking capability of the nonwoven material has been eliminated by movement and by increased extrusion of bentonite into the interface.

2.2 Internal Shear Strength Tests

2.2.1 Bentomat ST GCL--Laboratory direct shear tests also were conducted to determine the internal shear strength of the Bentomat ST GCL (Figure 4). The material samples used in the testing were obtained from GCL rolls at the project site and shipped to GeoSyntec in May 1995. The tests were conducted in accordance with ASTM D 5321

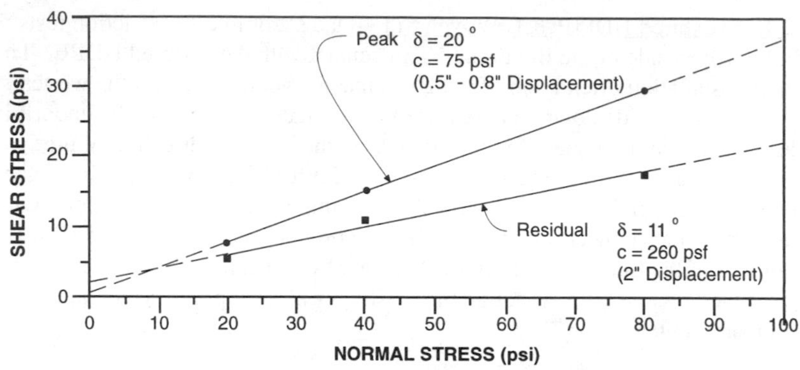

FIG. 2-- Textured HDPE/GCL (Woven Side) Interface Direct Shear Test Results

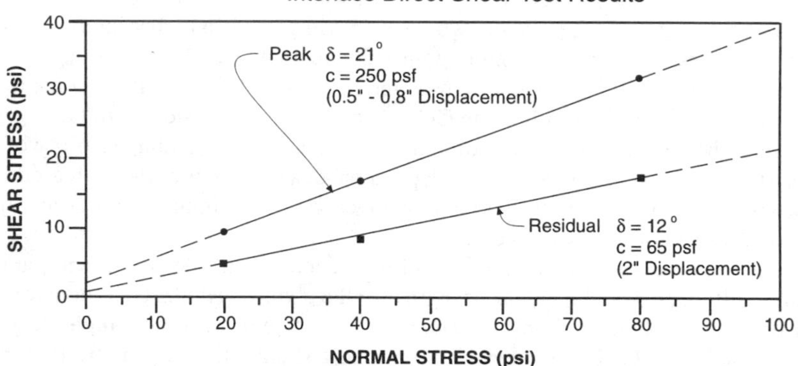

FIG. 3-- Textured HDPE/GCL (Nonwoven Side) Interface Direct Shear Test Results

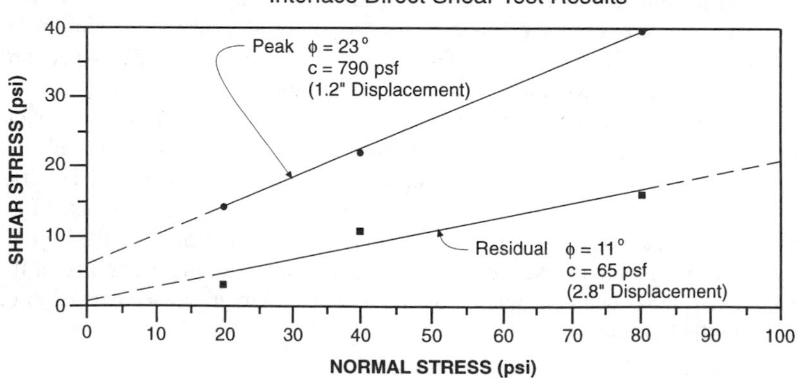

FIG. 4-- GCL Internal Shear Strength Direct Shear Test Results

Notes:
1) HDPE - Serrot 80-mil Single-Sided "Box and Point"
2) GCL - Bentomat ST
3) Hydration - 24 hours under 10 psi normal stress
4) Consolidation - 12 hours under each normal stress
5) Shear Rate - 0.04 in./min.
6) Sample Size - 12 in. X 12 in.
7) 1 Kpa = 0.145 psi
8) 1 Kpa = 20.875 psf
9) 1 cm = 0.394 in.

TABLE 1--Summary of interface test results (textured HDPE/Bentomat ST GCL)

Source and Materials	Test Date	Peak Strength	Residual Strength	Normal Stress Range, psi
Test Results-This Study				
GeoSyntec (Serrot "Box and Point" HDPE/GCL Woven)	April 1995	δ[3] = 20° c[4] = 75 psf	δ = 11° c = 260 psf	20-80
GeoSyntec (Serrot "Box and Point" HDPE/GCL Nonwoven)	April 1995	δ = 21° c = 250 psf	δ = 12° c = 65 psf	20-80
Test Results-Existing Data				
CETCO (Serrot "Box and Point" HDPE/GCL Woven)[1]	Nov. 1994	δ = 23° c = 52 psf	δ = 15° c = 69 psf	30-50
GeoSyntec (Serrot "Box and Point" HDPE/GCL Woven)[1]	Jan. 1995	δ = 12° c = 600 psf	δ = 5° c = 680 psf	25-100
Other HDPE/GCL Woven[2]	Various	NA	δ = 11° c = 100 psf	0.5-25
Other HDPE/GCL Woven[2]	Various	NA	δ = 5° c = 800 psf	25-80

[1] Recent test data supplied by manufacturer. Tests were run on 30-cm by 30-cm (12-inch by 12-inch), hydrated specimens.
[2] Average of values compiled from literature search and other CH2M HILL projects. All tests run on 30-cm by 30-cm (12-inch by 12-inch), hydrated specimens. Varying strength parameters presented for low and high normal stress ranges because of observed nonlinearity of shear strength envelope for available data.
[3] δ = interface friction angle
[4] c = cohesion intercept
[5] 1 kPa = 20.875 pounds per square foot (psf)
[6] 1 kPa = 0.145 psi

using test parameters selected to best model expected field conditions. Test conditions were similar to the conditions described previously for the interface friction tests, except that the HDPE material was not included in the test setup.

The shear rate used in the tests, 0.10 cm (0.04 inch)/minute, is relatively fast and is not expected to result in fully drained conditions within the specimens. Rates of approximately 0.00003 cm (0.000012 inch)/minute have been used to ensure fully drained conditions during shear [2 and 4]. There is relatively limited data at the present time to allow full evaluation of the impacts of shear rate on GCL internal strength. The limited data available [1, 2, and 4] suggest that GCL internal shear strength can be sensitive to the rate of shear and that decreased shear rates can result in decreased shear strengths in some cases (most likely as a result of dissipation of negative pore pressures). However, the degree of sensitivity appears to be dependent on the degree of hydration of the bentonite, the normal stress under which the specimen is sheared, whether peak or residual strengths are being measured, and other related factors. Additional research is required in this area.

Test results are presented graphically in Figure 4. Peak and residual internal shear strengths for the Bentomat ST GCL are summarized in Table 2. Also included in Table 2 are existing strength data for the Bentomat ST GCL for comparison purposes.

The residual GCL internal strength of $\phi = 11$ degrees and c = 3.1 kPa (65 psf) resulting from this study is comparable to typical values near 10 degrees measured for unreinforced bentonite with moisture contents greater than about 50 percent under fully drained conditions, where a shear rate of 0.00003 cm (0.000012 inch)/min was used [2 and 4]. Accordingly, use of a relatively fast shear rate for the tests in this study did not seem to have a significant impact on residual bentonite strength under hydrated and high normal stress conditions.

Hydration periods for the tests listed in Table 2 ranged from 24 to 72 hours, with hydration stresses ranging from nominal values (less than 7 kPa [1 psi]) up to the full normal stress under which the sample was sheared. Final moisture contents for the tested specimens ranged from approximately 100 percent to 200 percent, indicating some variability in the degree of hydration achieved. The test results listed in Table 2 are fairly consistent except for the peak strength from the August 1993 GeoSyntec test, where a higher ϕ (59 degrees) and a lower c (12.7 kPa [265 psf]) were measured relative to the other peak strength values reported in Table 2. The variation in c-ϕ combinations is expected to be due primarily to the difference in normal stress under which the tests were run. As noted in Table 2, the August 1993 test was run under very low normal stresses compared to the other tests. Similar c-ϕ combinations have been reported by others for reinforced Bentomat GCLs tested at low normal stresses [1]. Note that despite the high ϕ value of 59 degrees, the c-ϕ combination from the August 1993 test will result in a lower shear strength at low normal stresses (less than about 17 kPa [350 psf]) than will the peak c-ϕ combinations from the other tests listed in Table 2 that were run under higher normal stresses. The results shown in Table 2 are indicative of a nonlinear strength envelope, with a typically steeper shear stress versus normal stress curve at lower normal stresses. The nonlinearity of the shear strength envelope for GCLs has been documented by others [1, 3].

3.0 STABILITY ANALYSES

3.1 Design Basis

Using the material strength parameters determined from the testing program described above, static and seismic stability analyses were carried out to assess maximum allowable height of refuse and placement procedures. The analyses were conducted assuming that the open interior refuse face would be sloped at 3-to-1 (horizontal to vertical), and exterior refuse slopes would be sloped at no steeper than 3.5-to-1. The cell

TABLE 2--Summary of GCL internal shear strengths (Bentomat ST)

Source	Test Date	Peak Strength	Residual Strength	Normal Stress Range, psi
Test Results-This Study				
GeoSyntec	June 1995	ϕ [3] = 23° c = 790 psf	ϕ = 11° c = 65 psf	20-80
Test Results-Existing Data				
GeoSyntec[1]	August 1993	ϕ = 59° c = 265 psf	NA	0.3-3
GeoSyntec[1]	Jan. 1994	ϕ = 24° c = 700 psf	ϕ = 7° c = 515 psf	14-145
Other Data[2]	Various	NA	ϕ = 15° c = 100 psf	1.4-14
Other Data[2]	Various	NA	ϕ = 11° c = 250 psf	14-70

[1] Recent test data supplied by manufacturer. Tests were run on 12-inch by 12-inch, hydrated GCL specimens.
[2] Average of values compiled from literature search and other CH2M HILL projects. All tests run on 12-inch by 12-inch, hydrated GCL specimens. Varying strength parameters presented for low and high normal stress ranges because of observed nonlinearity of shear strength envelope for available data.
[3] ϕ = angle of internal friction
[4] 1 kPa = 20.875 psf
[5] 1 kPa = 0.145 psi

side slopes are inclined at 2.7-to-1 with 6.1-meter (20-foot) wide benches every 9.1 vertical meters (30 feet).

A plan view of the landfill is included as Figure 5, depicting the Cell 1 through Cell 5 layout. One of the representative sections used in the stability analysis, Section 1, is shown on Figure 6 and will be described for illustrative purposes. System components and material properties associated with Section 1 including refuse strength, are listed in Table 3. The selection of refuse strength can have a major impact on the stability evaluation. Design refuse strengths were assigned based on available studies in the literature [10 and 11]. Because refuse strength is difficult to determine and may vary over a wide range of c-ϕ combinations, parametric evaluations are recommended to assess analysis sensitivity to selected values. It should be noted that extensive stability analyses were run on numerous sections throughout the landfill to identify critical sections for further evaluation and to assess the sensitivity of the analyses to parameter variations.

Analyses were run using two separate strength cases for the critical liner components, as follows: (1) Residual strengths on both the side slopes and base, and (2) Residual strength on the side-slopes, peak strength on the base. Stark [5] advocates use of residual strengths on the side slopes and peak strengths on the base because of low induced shear stresses and deformations on the base compared to the side slopes. The side slopes are subject to downdrag shear stresses and associated deformation as a result

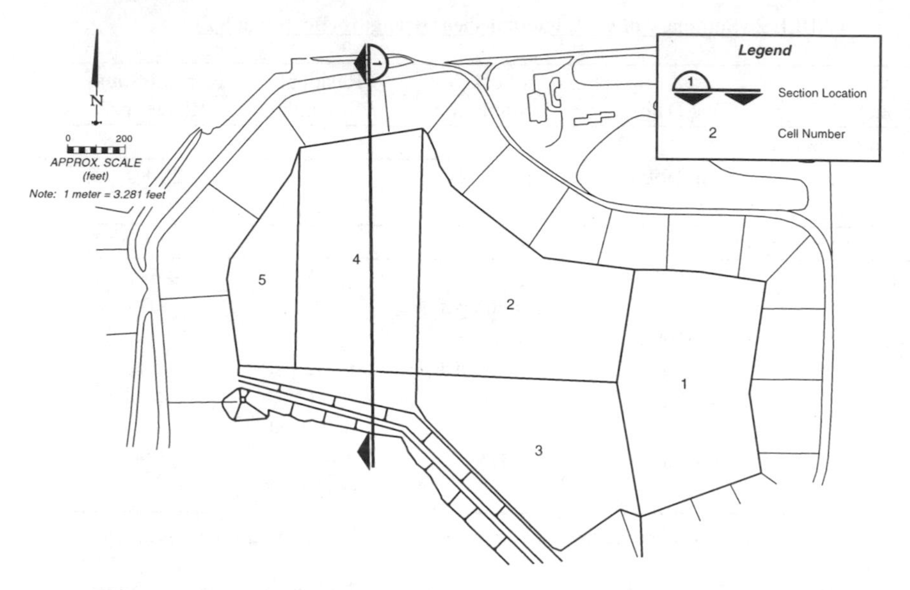

FIG. 5-- Plan View - Cells 1 Through 5, Anchorage Regional Landfill

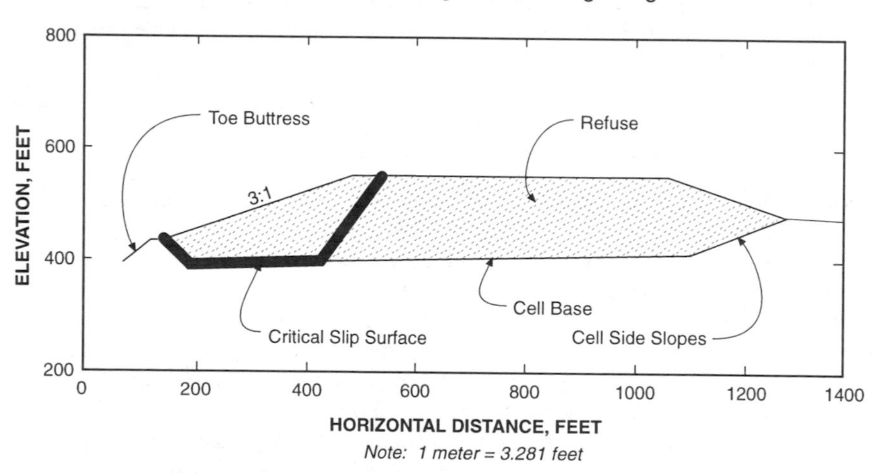

FIG. 6-- Elevation View - Section 1, Anchorage Regional Landfill

of waste settlement. The base is subject to much less induced stress. Safety factors for both strength cases were determined to aid assessment of acceptable stability. The strength of the side-slope lining system is controlled by the geotextile/HDPE interface where the peak and residual strength is represented by $\delta = 8°$. This interface is lower than even the residual strength of the underlying GCL under any stress conditions; accordingly, a predetermined slip plane is created above the geomembrane so that any

TABLE 3--System components and material properties
Section 1, Cells 4 and 5 Stability Analysis, Anchorage Regional Landfill

Location	System Components	Interface Friction Angle, Degrees	Angle of Internal Friction, Degrees	Cohesion, psf	Unit Weight, pcf
		Selected Material Properties			
Side Slopes	2' Sand Filter	NA	32	0	115
		30	NA	0	NA
	Reinforcing Geotextile				
		8	NA	0	NA
	80-mil HDPE FML (Single-Sided, Textured Dowı				
	Peak	20	NA	75	NA
	Residual	11	NA	260	NA
	GCL (Bentomat ST) Peak	NA	23	790	NA
	GCL (Bentomat ST) Residual	NA	11	65	NA
		26	NA	0	NA
	6" Sand Leveling Course	NA	34	0	120
Base	2' Sand Filter	NA	32	0	115
		29	NA	0	NA
	80-mil HDPE FML (Single-Sided, Textured Down)				
	Peak	20	NA	75	NA
	Residual	11	NA	260	NA
	GCL (Bentomat ST) Peak	NA	23	790	NA
	GCL (Bentomat ST) Residual	NA	11	65	NA
		26	NA	0	NA
	6" Sand Leveling Course	NA	34	0	120
	Refuse/Daily Cover Soil	NA	20	600	70
	Native Subgrade Soil (Dense Sand and Gravel)	NA	36	NA	135

Notes: (1) 1 kPa = 20.875 psf (2) 1 KN/m^3 = 6.37 pcf (3) NA = Not Applicable

slide will occur above the geomembrane and reduce the risk of damaging the HDPE/GCL lining system.

3.2 Static Analyses

Two-dimensional static stability analyses were performed on various landfill sections using the limiting equilibrium method. A state of limiting equilibrium on a selected slip surface is said to exist when the shear stress along the slip surface is equal to the available shear strength. The factor of safety is calculated as the actual available shear strength divided by the shearing stress along the selected slip surface. A computer model was developed with appropriate system geometry and material properties. Iterative analyses were conducted to determine the critical slip surface (lowest factor of safety) by using the program PCSTABL5 and the Simplified Jambu method of slices for noncircular surfaces [12]. A series of analyses was run for each critical section identified to assess the variation of factor of safety versus refuse height. The results of the static analyses for Section 1 are summarized in Table 4. The position of the critical slip surface determined for Section 1 is illustrated in Figure 6. The critical surface determined for Section 1 is typical of other sections analyzed and extends along the liner system to near

TABLE 4--Stability analysis results
(Cells 4 and 5 Anchorage Regional Landfill)

Section	Top of Refuse Elevation, m	Strength Case	Static Factor of Safety	Design Period, Years	Seismic Factor of Safety	Estimated Permanent Seismic Deformation, cm
1	168 (550 feet)	Residual (base & sides)	1.29	15	--	--
1	168 (550 feet)	Peak (base only)	1.72	15	1.10	< 2.5 (1 inch)

[1] Peak horizontal ground acceleration for 15-year design period = 0.31g
[2] Pseudostatic seismic coefficient = 0.155g

the refuse crest, then extends up through the waste to exit just beyond the crest. The maximum allowable refuse height for Section 1 is elevation 168 meters (550 feet) for operational reasons. At this height, static factors of safety of 1.29 and 1.72 were determined for residual and peak strength cases, respectively.

The allowable height of refuse placement in Cells 1 through 5, based on static stability considerations, was determined as the height resulting in a static factor of safety equal to 1.25 by using the residual strength case. A value of 1.25 was considered acceptable because of the temporary nature of the configurations analyzed, the use of residual strengths on the landfill base and side slopes, and the anticipated conservatism in the selected refuse properties [10, 11]. Determination of adequate stability also was based on the results of the seismic analyses as described below.

3.3 Seismic Analysis

The first step in assessing the seismic stability of refuse in the cells was to select appropriate earthquake design criteria. Subtitle D of the EPA Landfill Regulations [13] specifies that all landfill containment structures be designed to resist the maximum expected horizontal acceleration with a 90 percent or greater probability of not being exceeded over a design period of 250 years. This criterion applies to the final landfill configuration.

The sections analyzed for Cells 1 through 5 represent a temporary construction case which will cease to be a concern when adjacent cells are filled due to buttressing effects. Accordingly, much lower design periods can be associated with this temporary construction case compared to the 250-year design period associated with the final landfill configuration. Design periods for each of the sections analyzed were determined from the landfill development plan. Design periods were selected using the basic assumption that refuse would be placed instantaneously at the beginning of the filling periods.

To design for the same level of risk as specified in Subtitle D, peak horizontal ground accelerations with a 90 percent probability of not being exceeded over the selected design periods were determined from a site-specific seismic risk assessment [14]. Peak horizontal acceleration values ranging from 0.20g to 0.31g were determined for the design periods in question. The design period and associated peak acceleration value for Section 1 are listed in Table 4.

Section 1 was initially analyzed for seismic stability using a pseudostatic limiting equilibrium approach. A pseudostatic seismic coefficient equal to 1/2 the peak horizontal acceleration was selected for analysis [11]. Both peak and pseudostatic acceleration values are noted in Table 4 along with the resulting seismic factor of safety for refuse placed to the recommended maximum elevation for Section 1. Factors of safety near 1.0 or greater indicate that permanent deformations, if any occur, will be relatively small under the design seismic loading [11].

The pseudostatic analysis described above provides only a general indication of slope behavior under seismic loading. To further assess slope behavior, permanent deformations under earthquake loading were estimated using the approach suggested by Seed and Makdisi [15]. Their approach is based on Newmark's original work [16]. Because there is little information regarding attenuation or amplification characteristics of refuse, the analyses for this project assumed that the effective peak acceleration in the refuse was constant and equal to the peak acceleration at the base of the landfill. Predicted deformation under the seismic design load for Section 1 was less than 2.5 cm (1 inch) under the peak strength case, as shown in Table 1. Because the calculated seismic displacement was consistent with the small displacements required to mobilize the peak strengths in the laboratory direct shear tests (see Figures 2 through 4), the peak strength case was deemed appropriate for design of this interim configuration under seismic loading. Other sections were similarly analyzed. The relatively low deformation determined from this analysis was considered to pose relatively small risk to the planned landfill lining system. Maximum allowable deformations of 15 to 30 cm (0.5 to 1 foot) have typically been used in practice for design of geosynthetic liner systems [11].

4.0 QUALITY ASSURANCE AND CONFORMANCE TESTING PROGRAM

One significant advantage of a GCL over a compacted clay liner (CCL) is the relative simplicity of quality assurance and conformance testing. However, well-written specifications detailing a thorough quality assurance program and conformance testing during construction are necessary to ensure the adequacy of the installed GCL. The specifications for the Cells 4 and 5 project required quality assurance and conformance testing through the following three phases of construction:

1. Prior to GCL Fabrication–Information was requested describing the manufacturer's product and documenting that the proposed GCL material conforms to the project specifications. Typically, standard manufacturer's literature comprises much of this submittal for initial material approval. Additional testing may be required at this stage if test data are unavailable for project-specific test parameters. The manufacturer's quality assurance/quality control plan and material samples also were requested at this time.

2. During GCL Fabrication–Factory test results on representative rolls of the material as it is manufactured for the project were required. Parameters for strength and permeability tests were reviewed and preapproved by the project engineer to ensure that

field conditions were accurately modeled. Factory material samples were submitted to the engineer for independent testing to confirm and/or supplement the contractor's data.

3. During Construction–Conformance tests were run on samples of the material delivered to the site. Tests can be run on any properties that may be of particular importance on a given project.

Requirements for quality assurance and conformance testing on GCLs for the three project phases described above are summarized in Table 5 for the ARL Cells 4 and 5 project. More detailed considerations of quality assurance and conformance testing of GCLs are available in Reference [17].

TABLE 5--GCL Quality assurance and conformance testing program (Cells 4 and 5, Anchorage Regional Landfill)

Project Phase	Information Required	Testing Requirements	Testing Frequency	Comments
Prior to GCL manufacture	1) Material data-typically manufacturer's standard literature	Additional strength or permeability testing may be required to demonstrate compliance with project-specific test conditions.	As required	...
	2) Manufacturer's QA/QC program			
	3) GCL material samples			
During GCL manufacture	1) Factory test results	Specified strength and material properties (by manufacturer).	1 test per 50,000 ft^2	Require engineer's preapproval of test procedures to best model field conditions.
		Permeability (by manufacturer).	1 test per 250,000 ft^2	Require engineer's preapproval of test procedures to best model field conditions.
	2) Factory samples	Strength and permeability (by engineer).	As required	Independent testing to confirm or supplement manufacturer's data.
During construction	1) Conformance testing on materials delivered to site	Mass/area	1 test per 50,000 ft^2	Contractor's responsibility-conducted by independent third-party lab.
		Bentonite content	1 test per 50,000 ft^2	"
		Free swell	1 test per 50,000 ft^2	"
		Grab tensile and elongation	1 test per 50,000 ft^2	"

[1] 1 m^2 = 10.765 ft^2

The results of index property conformance testing on GCL samples taken at the project site are summarized in Table 6. The results indicate a fair degree of variability in the manufactured product, although only a relative few individual test specimens failed to meet the specified values.

TABLE 6--Bentomat ST GCL conformance testing results
(Cells 4 and 5, Anchorage Municipal Landfill)

Test	Test Designation	Unit	Number of Tests	Specified Minimum Values	Average	Maxi-mum	Mini-mum	Standard Deviation
Mass/unit area	ASTM D 3776	lbs/yd^2 (4)	84	9.5	10.2	12.6	9.2	0.79
Bentonite content (1)	ASTM D 3776	psf (5)	84	0.9	1.1	1.3	0.9	0.09
Free swell	USP-NF-XVII [18]	mLs	5	24	56.0	60.0	50.0	5.48
Grab strength-MD (2)	ASTM D 4632	lb(6)	280	90	129.1	253.0	81.0	21.31
Grab strength-TD (3)	ASTM D 4632	lb	279	90	203.6	362.0	122.0	39.28
Grab elongation-MD	ASTM D 4632	%	280	15	16.7	31.0	11.0	2.47
Grab elongation-TD	ASTM D 4632	%	280	15	114.3	260.0	14.0	42.30

1 Bentonite content at 12% moisture content
2 MD = machine direction
3 TD = transverse direction
4 1 kPa = 187.875 lbs/yd^2
5 1 kPa = 20.875 psf
6 4.45 N = 1 lb

Of particular interest in the results is the relatively high degree of variability in the grab strength and grab elongation values. The following observations can be made:

- Bentomat ST is made up of both nonwoven and woven geotextiles which are needle-punched together. The two geotextiles have widely varying strength and elongation properties which, when tested together as in a GCL, can result in the wide variations in results observed.
- The tensile force versus displacement curve for a GCL composed of two different geotextiles typically has two peaks, one representing the woven geotextile and one representing the nonwoven geotextile. CETCO's recommended test procedure is to run the test out to a sufficiently large displacement so that both peaks occur, and then use the higher peak as the grab strength [19]. The widely varying grab strengths observed in the above test results are expected to have resulted from woven peak strengths being recorded as the grab strength for some samples and nonwoven peak strengths recorded for other samples. The first peak could represent either the woven or nonwoven geotextile, depending on material characteristics and test orientation. The widely varying elongation values recorded also indicate that both geotextile types influenced the test results, with

values ranging from 11 percent (characteristic of a woven fabric) up to 260 percent (more characteristic of a nonwoven).

- The grab strengths in the transverse direction are significantly higher than the grab strengths in the machine direction. This is probably attributable to the typically higher strength of nonwoven geotextiles in the transverse direction compared to the machine direction.

5.0 RESULTS AND CONCLUSIONS

The following conclusions are based on the testing and evaluations conducted during the course of the ARL Cells 4 and 5 project:

1. Project-specific strength testing should be carried out on the GCL material proposed for any given application. Testing of the GCL internal shear strength as well as interface friction of anticipated low-strength interfaces should be carried out. Test parameters should model expected field conditions, including degree and rate of hydration, degree and rate of consolidation, applied normal stresses under which hydration or consolidation may occur, and loading rate.

2. Interface and internal shear strength tests should be carried out under the range of stresses expected in the field, since failure envelopes are typically nonlinear. Additionally, different interfaces or materials may control design (provide the lowest shear strength) at different stress conditions.

3. For the Serrot 80-mil single-sided, textured "box and point" HDPE and Bentomat ST GCL tested here, results were as follows:

- Interface strengths between the textured HDPE and the woven or nonwoven side of the GCL were not significantly different.
- The interface strength between the HDPE and the GCL was less than the GCL internal strength at relatively small displacements (peak conditions); the GCL internal strength was less than the interface strength at relatively large displacements (residual conditions).
- Interface peak strengths were mobilized at displacements of 1.3 to 2.0 cm (0.5 to 0.8 inch); GCL internal peak strengths were mobilized at a displacement of about 3.0 cm (1.2 inch).
- Interface residual strengths were mobilized at displacements of about 5.1 cm (2.0 inches); GCL internal residual strengths were mobilized at a displacement of about 7.1 cm (2.8 inches). The greater required displacements for full development of the GCL peak and residual strengths are probably attributable to the needle-punched reinforcement in the GCL.
- Interface strength results between the Serrot "box and point" HDPE and GCL were generally similar to values measured for other types of textured HDPE geomembranes.

4. The low residual internal GCL strength measured in this study ($\phi = 11°$, c = 3.1 kPa [65 psf]) compares well with strength values for unreinforced hydrated bentonite sheared under fully drained conditions. Accordingly, the relatively fast shear rate used for this study does not appear to have significantly impacted the test results.

5. This project demonstrates that use of a side-slope woven reinforcing geotextile above the HDPE can result in a predetermined slip plane which will control the design, even under residual conditions in the GCL. Even the residual internal strength of the GCL measured ($\phi = 11°$, c = 3.1 kPa [65 psf] under high normal stresses) was higher than the $\delta = 8°$ anticipated for the smooth HDPE/woven geotextile interface. The predetermined slip plane above the HDPE results in cover material sliding on the HDPE if any movement does occur, effectively isolating the HDPE and GCL and reducing the risk of damage to the lining system.

6. Acceptable factors of safety for slope stability analyses should be selected with due consideration given to whether peak or residual strengths are used in the analyses. Acceptance of lower factors of safety may be appropriate if residual strengths are used.

7. To design for the same level of risk specified in EPA Subtitle D for the final landfill configuration, seismic stability analyses of interim landfill configurations can be conducted using a reduced design earthquake acceleration corresponding to the interim period under consideration.

8. If the factor of safety for landfill stability under design seismic loading is near 1.0 or less, seismic deformation analyses can be carried out to estimate the magnitude of movement and assess the potential impact of such movement.

9. The deformations estimated from the seismic analysis should be consistent with the deformations required in the laboratory tests to mobilize the strength values used in the analysis.

10. Quality assurance and conformance testing of GCLs is important and should cover all project phases, including before and during GCL fabrication, and during construction as the material is brought to the site.

11. Results of index property conformance testing on Bentomat ST GCL samples indicate some variability in the manufactured product, although few individual test specimens failed to meet the specification requirements. The wide variability observed in the grab strength and elongation test results is primarily a function of testing procedure on the composite GCL material. Test procedures need to be standardized to provide more meaningful results with minimal variability.

ACKNOWLEDGMENTS

The testing and evaluations described here were funded by the Municipality of Anchorage Solid Waste Services department. We extend our sincere appreciation to Solid Waste Services' project representatives Joel Grunwaldt and Bill Kryger for their cooperation throughout this interesting project. The assistance of representatives of the geosynthetic manufacturers is also greatly appreciated, particularly Rick Taylor of Serrot and Robert Trauger of CETCO. The writers also wish to thank Larry Well and Mike Reimbold of CH2M HILL for their input during preparation of this paper, and Robert Swan of GeoSyntec Consultants for his assistance during the laboratory testing program.

REFERENCES

[1] Daniel, D.E. and Gilbert, R.B., "Geosynthetic Clay Liners for Waste Containment and Pollution Prevention," Continuing Education Course, University of Texas at Austin College of Engineering, October 6-7, 1993.
[2] Daniel, D.E., Shan, H.Y., and Anderson, J.D., "Effects of Partial Wetting on the Performance of the Bentonite Component of a Geosynthetic Clay Liner," Geosynthetics '93, Vol. 3, Industrial Fabrics Association International, St. Paul, Minn., pp 1483-1496.
[3] Daniel, D.E., "Geosynthetic Clay Liners in Landfill Covers," 31st Annual Solid Waste Exposition, Solid Waste Association of North America, San Jose, Calif., August 1993.
[4] Shan, H.Y. and Daniel, D.E., "Results of Laboratory Tests on a Geotextile/Bentonite Liner Material," Geosynthetics '91, Vol. 2, Industrial Fabrics Association International, St. Paul, Minn., pp 517-535.
[5] Stark, T.D. and Poeppel, A.R, "Landfill Liner Interface Strengths from Torsional-Ring-Shear Tests," Journal of Geotechnical Engineering, Vol. 120, No. 3, March 1994.
[6] Derian, L., Gharios, K.M., Kavazanjian, E., Jr., and Snow, M.S., "Geosynthetics Conquer the Landfill Law," Civil Engineering, December 1993.
[7] Geotechnical Fabrics Report, "Performance of a Geosynthetic Liner System in the Northridge Earthquake," March 1994.
[8] GeoSyntec Consultants, "Final Report, Direct Shear Testing, Anchorage Regional Landfill, Cells 4 and 5," September 28, 1995.
[9] American Society for Testing and Materials (ASTM), "D5321-92, Standard Test Method for Determining the Coefficient of Soil and Geosynthetic or Geosynthetic and Geosynthetic Friction by the Direct Shear Method." 5 pp.
[10] Singh, S. and Murphy, B., "Evaluation of the Stability of Sanitary Landfills," Symposium on Geotechnics of Waste Fills - Theory and Practice, ASTM STP 1070, Philadelphia, Pa., 1990.
[11] Richardson, Gregory, and Kavazanjian, Edward, "RCRA Subtitle D (258) Seismic Design Guidance for Municipal Solid Waste Landfill Facilities," U.S. Department of Commerce, April 1995.
[12] Purdue University, School of Engineering, "PCTABL5 User Manual."
[13] Code of Federal Regulations, Title 40. Part 258.14–"Criteria for Municipal Solid Waste Landfills, Seismic Impact Zones," Effective October 9, 1993.
[14] Earth Mechanics, Inc., "Seismic Hazard Evaluation, Anchorage Regional Landfill," April 15, 1994.
[15] Seed, H.B. and Makdisi, "A Simplified Procedure for Estimating Earthquake Induced Deformation in Dams and Embankments," UCB/EERC 77/19, August 1977.
[16] Newmark, N.M., "Effects of Earthquakes on Dams and Embankments," Geotechnique, Vol. 15, No. 2, 1965 pp 139-160.
[17] Daniel, D.E. and Koerner, R.M., "Quality Assurance and Quality Control For Waste Containment Facilities," EPA Technical Guidance Document, July 1993.
[18] U.S. Pharmacopeia—National Formulary XVII, "USP-NF-XVII, Test Method for the Free Swell of Bentonite Clay," Page 1210.
[19] Personal Communication, Robert Trauger, CETCO, December 1995.

John R. Siebken[1], Robert H. Swan, Jr.[2], and Zehong Yuan[3]

SHORT-TERM AND CREEP SHEAR CHARACTERISTICS OF A NEEDLEPUNCHED THERMALLY LOCKED GEOSYNTHETIC CLAY LINER

REFERENCE: Siebken, J. R., Swan, Jr., R. H., and Yuan, Z., **"Short-Term and Creep Shear Characteristics of a Needlepunched Thermally Locked Geosynthetic Clay Liner,"** *Testing and Acceptance Criteria for Geosynthetic Clay Liners, ASTM STP 1308,* Larry W. Well, Ed., American Society for Testing and Materials, 1997.

ABSTRACT: A series of constant-rate direct shear tests were conducted on a needlepunched thermally locked geosynthetic clay liner (GCL) in accordance with ASTM Test Method for Determining the Coefficient of Soil and Geosynthetic or Geosynthetic and Geosynthetic Friction by the Direct Shear Method (D 5321). The test results demonstrate that the needlepunched thermally locked reinforcing fibers provide substantial short-term shear strength to a GCL. However, there is a growing concern that the long-term shear strength of this type of GCL can be affected due to the potential of creep within the reinforcing fibers under sustained constant loads which occur in the field. An attempt was made to address this concern through an incrementally-loaded creep shear test conducted in a newly developed constant-load (creep) shear testing device. The results of the creep shear test to date show that the GCL has undergone relatively small shear displacements with incremental shear rates decreasing with time within each loading phase.

KEYWORDS: geosynthetic clay liner, shear strength, creep shear, direct shear, needle-punched, thermally locked, reinforced.

---

[1]Supervisor of Technical Services, National Seal Company, 1255 Monmouth Blvd., Galesburg, IL 61401.

[2]Laboratory Manager and Assistant Program Manager, respectively, GeoSyntec Consultants, Soil-Geosynthetic Interaction Testing Laboratory, 5775 Peachtree Dunwoody Rd., Suite 11D, Atlanta, GA 30342

Geosynthetic clay liners (GCLs) are being used with increasing frequency in landfill lining and closure systems. A primary concern of GCLs used in these applications is the ability to withstand and transmit the shear stresses imposed during their service life. The adequacy of the internal strength of the GCL must be addressed for each application via laboratory tests. One of these laboratory tests is the constant-rate direct shear test (ASTM D 5321) which is typically used to evaluate the short-term shear strength of GCLs and interfaces between GCLs and other materials (i.e., soil and/or other geosynthetics). Another type of test is the constant-load creep shear test which was developed by the authors to evaluate the long-term creep behavior of GCLs and GCL interfaces.

This paper focuses on both the short-term and long-term behavior of a needle-punched thermally locked GCL under normal stress conditions that typically occur in a landfill lining system. First, a series of constant-rate direct shear tests were performed on the thermally locked GCL to define its short-term shear strength behavior. Then a constant-load creep shear test consisting of four loading increments was performed on the thermally locked GCL specimen using a large scale creep shear test device, which was developed specifically for this research study. The test conditions for the constant-rate direct shear and constant-load creep shear tests are described in detail in this paper. The results of the two types of the direct shear tests are presented, compared, and discussed.

CONSTANT-RATE DIRECT SHEAR TESTS

Equipment

The constant-rate direct shear tests were performed in general accordance with the ASTM D 5321. The tests were conducted in a large direct shear device which was developed by the authors. The direct shear test device contained an upper and lower shear box. The upper shear box measured 300 mm by 300 mm in plan and 75 mm in depth. The lower shear box measured 300 mm by 350 mm in plan and 75 mm in depth. Each test specimen was set up and tested under the specific conditions as described below.

Configuration of Test Specimen

A series of constant-rate direct shear tests was conducted to evaluate the short-term shear strength behavior of the thermally locked GCL. The test series consisted of seven tests with each test conducted at a different level of normal stresses ranging from 35 to 670 kPa. The configuration of the test specimens used in the test series, from top to bottom, was as follows:

- rigid substrate with textured steel gripping surface.
- thermally locked GCL; and
- rigid substrate with textured steel gripping surface.

It is noted that the textured steel gripping surfaces were developed by the authors during the course of establishing the appropriate testing procedures for measuring short-term internal shear strength of GCLs [1]. The textured steel gripping surfaces were employed to minimize slippage between the geotextile components of the GCL and the rigid wooden substrate, therefore providing a relatively uniform transfer of shear load into the GCL specimen.

Test Procedures and Conditions

A fresh GCL specimen was trimmed from the bulk sample of the thermally locked GCL. Each test was conducted specifically to evaluate the internal strength of the thermally locked GCL specimen. This was achieved by constraining the thermally locked GCL specimen such that shearing could only occur through the bentonite component of the GCL specimen. Specimen constraint was accomplished by completely bonding the geotextile component on each side of the thermally locked GCL specimen to the rigid wooden substrate with the use of textured steel gripping surfaces. The ends of each geotextile were then sandwiched between a second rigid wooden substrate prior to testing as shown in Fig. 1. The entire test specimen was then placed in the shear box to provide confinement for the exposed bentonite component.

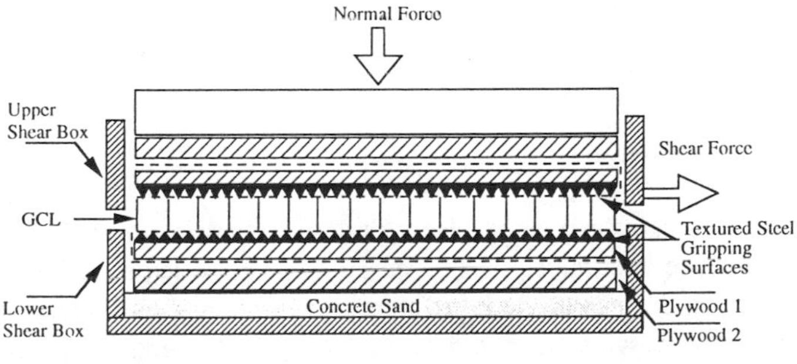

FIG. 1--Schematic of GCL Test Specimen Configuration.

For each test, the entire test specimen was soaked in tap water for 24 hours under each applied normal stress prior to being sheared. The normal stress was applied to the test specimen through an air bladder system prior to submerging the thermally locked GCL in tap water. After the 24-hour soaking period, the test specimen was sheared immediately by displacing the upper shear box at a constant displacement rate of 1 mm /min. Shearing of the test specimen was continued until a total shear displacement of 50 mm was achieved.

For each test, shear loads and shear displacements were measured by an electronic load cell and a linear variable differential transformer (LVDT), respectively. The shear load and displacement data were recorded by a computer data acquisition system through the test.

Test Results

The test data from each constant-rate direct shear test were plotted on a graph of shear force versus horizontal displacement as shown in Fig. 2. The peak value of shear force was used to calculate the peak shear strength. The residual shear strength was calculated using the shear force measured at a shear displacement of 50 mm. It is noted that a constant shearing area of 0.1 m^2 was used when computing normal and shear stresses. This area was the initial plan area of the thermally locked GCL test specimen used in each test.

The calculated shear strengths were plotted on a graph of shear stress versus normal stress and the results were used to evaluate total-stress peak and residual strength envelopes. A best-fit straight line was drawn through the data points from the test series to obtain total-stress peak and residual friction angles and adhesions. The plot of shear stress versus normal stress for the test series is shown in Figure 3. The friction angle and adhesion values derived from the plotted test results are also indicated in Fig. 3.

CONSTANT-LOAD CREEP SHEAR TEST

Description of Equipment

A large-scale creep shear test device was designed and fabricated by the authors in general accordance with the equipment descriptions provided in ASTM D 5321. The creep shear test device was designed to evaluate the long-term creep shear behavior of various geosynthetic materials, including GCLs, under various normal and shear stress conditions, typical of those found in landfill lining and closure systems. The creep shear test device consisted of five major components as described below:

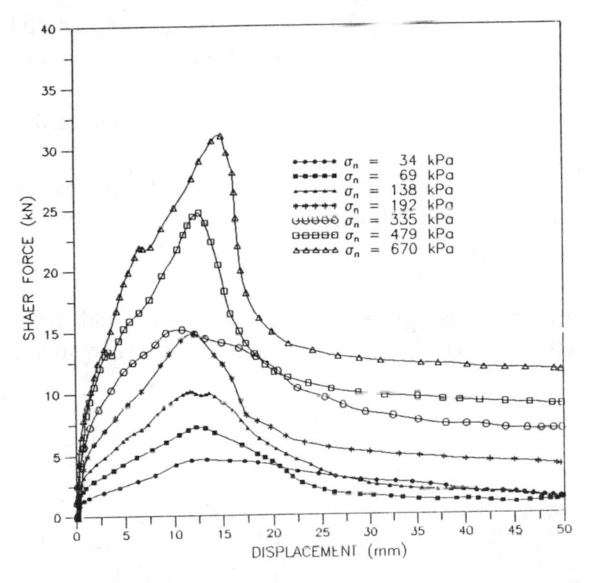

FIG. 2--Plot of shear force versus displacement for the thermally locked GCL.

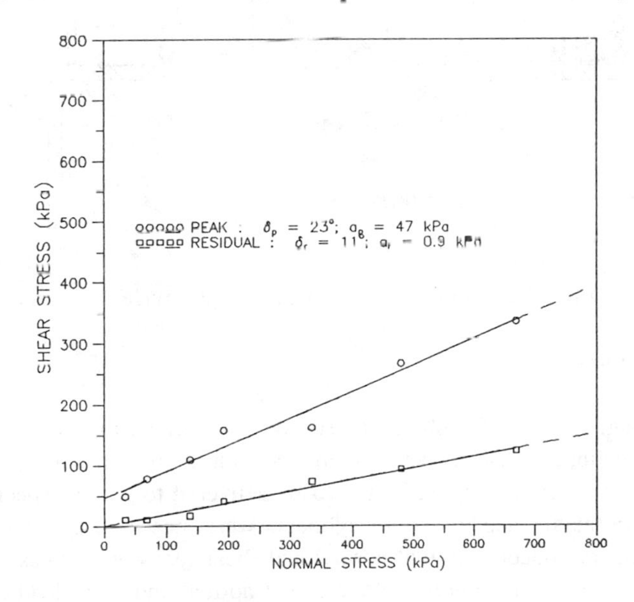

FIG. 3--Plot of shear stress versus normal stress for the thermally locked GCL.

- a rigid supporting table;
- a upper shear box measuring 300 mm by 300 mm in plan and 75 mm in depth and a lower shear box measuring 300 mm by 350 mm in plan and 75 mm in depth;
- a containment box to allow for fully submerged testing of the test specimens;
- an air bladder system for applying normal loads to the test specimen; and
- three parallel-connected 150 mm diameter air cylinders for applying shear loads to the test specimen.

The creep shear test device was capable of delivering normal and shear stresses ranging from 7 to 840 kPa. The creep shear test device is schematically shown in Fig. 4.

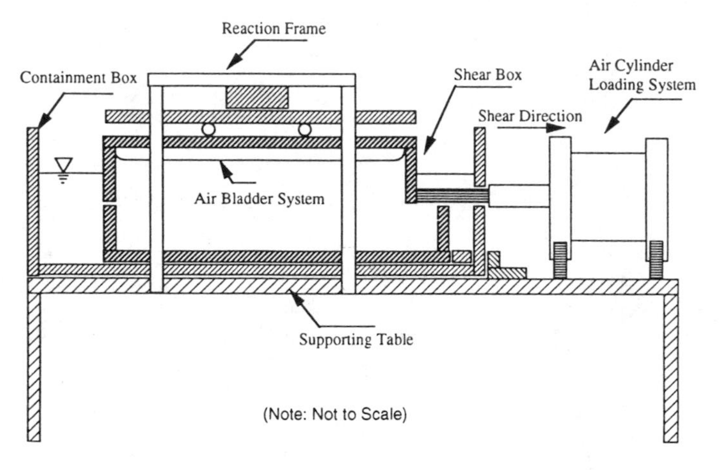

FIG. 4--Schematic of creep shear test device.

Equipment Calibration

Prior to testing, a series of calibration tests were performed to establish the relationship (i.e., calibration curve) between applied source pressures (i.e., pressure in the air bladder or air cylinders) and actual loads delivered to the test specimen. Source pressures in the air bladder and air cylinders were measured by a variable reluctance pressure transducer. Actual normal and shear loads were measured by an electronic load cell. The source pressure and actual normal and shear load data were recorded by a computer data acquisition system.

For the calibration of the air bladder system, the load cell was placed between the air bladder system and the rigid steel reaction frame (see Fig. 4). A normal load versus pressure curve was established for source pressures ranging from 70 to 840 kPa. The maximum load delivered by the air bladder was approximately 80 kN at a source pressure of 840 kPa. It is noted that the actual normal load on top of the air bladder system is not the same as that transmitted to the test specimen. The actual applied normal load to the test specimen equals the sum of the measured reaction load, weight of the air bladder system, and the test specimen itself.

For the calibration of the shear loading system, the electronic load cell was mounted to the front of the shear box and connected to the air cylinders. A shear load versus pressure curve was established for source pressures ranging from 7 to 840 kPa. The maximum load delivered by the air cylinder loading system was approximately 80 kN at a source pressure of 840 kPa. Since the loading system was direct (no mechanical advantage), it was assumed that the applied force would remain constant independent of displacement.

To further confirm that the creep shear test device functioned properly, a trial test series was conducted to evaluate the peak interface shear strength between a nonwoven geotextile and 1.5-mm thick smooth high density polyethylene (HDPE) geomembrane. The tests were conducted in accordance with the test procedures as follows:

- a fresh specimen of concrete sand was compacted into the lower shear box by hand tamping to a relatively dense state under dry conditions forming a 75-mm thick bedding layer;
- fresh specimens of the geotextile and geomembrane were attached to the upper and lower shear boxes, respectively, with mechanical compression clamps;
- a rigid substrate was then placed on top of the geotextile specimen;
- a normal stress was applied to the test specimen through the air bladder system; and
- shear loads were then applied to the test specimen through the air cylinder loading system. The shear load was progressively increased by increasing the air pressure within the air cylinders until sliding occurred at the test interface. The maximum pressure within the air cylinders was recorded.

The results of the trial tests indicated a peak friction angle of 12 degrees and an adhesion of 2.0 kPa for the geotextile-smooth geomembrane interface under normal stresses ranging from 70 to 700 kPa. The results are consistent with those measured using a constant rate of displacement direct shear testing device.

Test Procedures and Conditions

The configuration of the creep shear test specimen was the same as that shown in Fig. 1. The test specimen was constructed in the same way as previously described for the constant-rate direct shear testing. The thermally locked GCL specimen was subjected to test conditions as follows:

- Soaking: soaked in tap water for 72 hours under a normal stress of 414 kPa. The normal stress used for soaking was applied prior to immersion;
- Consolidation and Shearing Phase 1: consolidated for 120 hours under a normal stress of 414 kPa and then subjected to a constant shear stress of 207 kPa for 500 hours;
- Consolidation and Shearing Phase 2: consolidated for 120 hours under a normal stress of 483 kPa and then subjected to a constant shear stress of 242 kPa for 500 hours;
- Consolidation and Shearing Phase 3: consolidated for 120 hours under a normal stress of 552 kPa and then subjected to a constant shear stress of 276 kPa for 500 hours;
- Consolidation and Shearing Phase 4: consolidated for 120 hours under a normal stress of 621 kPa and then subjected to a constant shear stress of 311 kPa for 500 hours;
- For each phase, the target shear load was applied to the thermally locked GCL specimen within approximately 10 to 15 seconds in a controlled manner; and
- Vertical displacements of the test specimen were measured by two dial gages attached to the top of the GCL specimen in the upper shear box. Shear displacements of the test specimen were measured by a dial gage attached to the upper shear box.

The ratio of shear load to normal load used in each loading phase was 1:2 and was selected to simulate the loading conditions of a landfill lining system constructed on a 2H:1V side slope. The loading path used in the constant-load creep shear test is shown in Fig. 5 together with the short-term peak shear strength envelope of the thermally locked GCL. As it can be seen from Fig. 5, the applied shear stress varied from approximately 90 to 99 percent of the short-term peak strength of the thermally locked GCL. It is noted that the intent of this test was to fail the thermally locked GCL by progressively increasing the shear stress level (percentage of the short-term peak strength) with each load increment. If the GCL did not fail within the 500 hour load increment, the shear stress level was to be further increased. However, the 1:2 load path was to be maintained with each load increment. The test is still on going and additional load increments will be applied until failure hopefully occurs.

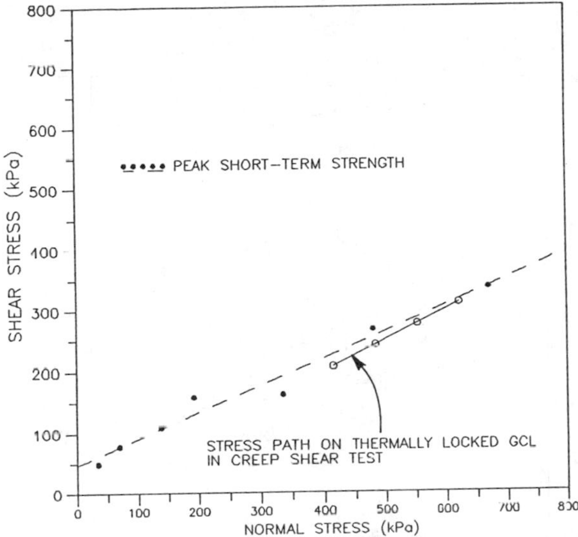

FIG. 5-- Plot of the loading path for the constant-load creep shear test on the thermally locked GCL with the short-term peak shear strength envelope of the GCL.

Test Results

A creep shear test consisting of four load increments was conducted on the thermally locked GCL. For each load increment of the test, the thermally locked GCL was first consolidated for 120 hours under each applied normal stress and then subjected to a constant shear load for 500 hours. The results of a typical loading phase are graphically presented in Figs. 6 through 8. Fig. 6 indicates that the thermally locked GCL was compressed under the soaking/consolidation normal stress. Fig. 7 indicates that the thermally locked GCL underwent a shear distortion immediately after application of the shear load and then started to "creep" (i.e., develop shear displacement with time under the applied constant shear load). Fig. 7 also indicates that there was very little change in the thickness of the thermally locked GCL during shearing. Fig. 8 indicates that incremental shear rates continued to decrease throughout shearing for each load increment.

A summary of shear displacement and incremental shear rates at select times is presented in Table 1. This table presents the total shear displacements and incremental shear rates at 1, 10, 100, and 500 hours of elapsed time for each phase of the creep shear test. It is noted that the values of total displacement and incremental shear rate at these selected times presented in Table 1 were derived by linear interpolation between the actually measured data points at each specific time period of interest.

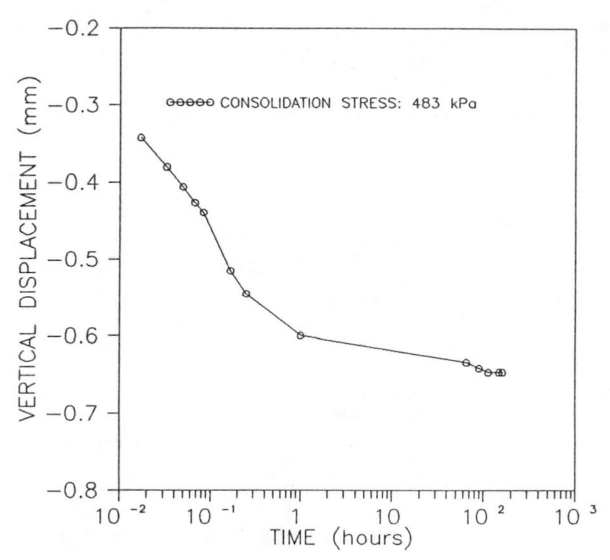

FIG. 6--Plot of vertical displacement versus log time for the thermally locked GCL.

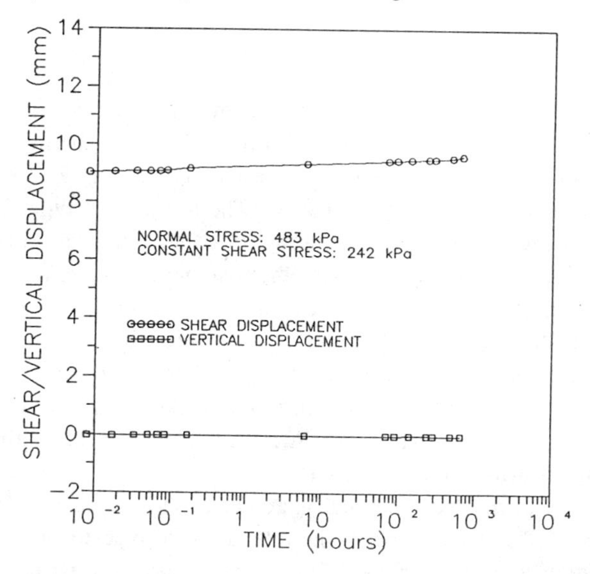

FIG. 7-- Plot of shear and vertical displacements versus log time for the thermally locked GCL.

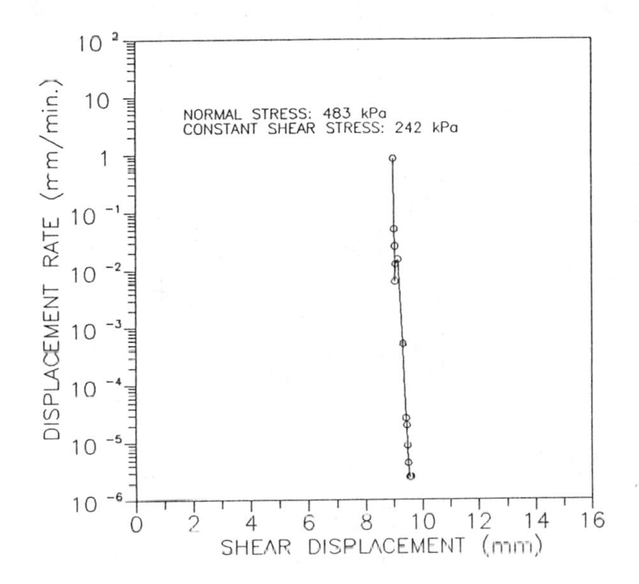

FIG. 8-- Plot of log displacement rate versus shear displacement for the thermally locked GCL.

SUMMARY AND DISCUSSIONS

A series of constant-rate direct shear tests were conducted on the thermally locked GCL under hydrated conditions. The results indicated a peak friction angle of 23 degrees and an peak adhesion of 47 kPa, and a residual friction angle of 11 degrees and a residual adhesion of 0.9 kPa under a series of normal stresses ranging from 34 to 670 kPa. The thermally locked GCL mobilized its peak strength at a shear displacement of approximately 12 mm and then decreased rapidly as displacement increased. The decrease of shear strength after the peak is primarily due to rupture of the reinforcing fibers. The thermally locked GCL finally achieved its residual shear strength when all of the reinforcing fibers ruptured. However, it is noted that the reported residual strengths in this paper were measured at a shear displacement of 50 mm.

TABLE 1--Summary of constant-load creep shear test results.

Test Specimen	Elapsed Time (hours)	Phase I (σ_n = 414 kPa, τ = 207 kPa)		Phase II (σ_n = 483 kPa, τ = 242 kPa)		Phase III (σ_n = 552 kPa. τ = 276 kPa)		Phase IV (σ_n = 621 kPa, τ = 311 kPa)	
		Total Shear Displacement (mm)	Incremental Shear Rate (mm/min)	Total Shear Displacement (mm)	Incremental Shear Rate (mm/min)	Total Shear Displacement (mm)	Incremental Shear Rate (mm/min)	Total Shear Displacement (mm)	Incremental Shear Rate (mm/min)
Internal Strength of	1	7.2	1.37×10^{-2}	9.2	1.31×10^{-2}	10.3	3.35×10^{-3}	11.1	2.15×10^{-3}
Needle-Punched Thermally	10	7.9	2.52×10^{-3}	9.3	4.78×10^{-4}	10.3	1.51×10^{-3}	11.2	2.33×10^{-5}
Locked GCL Under	100	8.2	2.95×10^{-5}	9.4	1.79×10^{-5}	10.6	5.58×10^{-5}	11.3	8.94×10^{-6}
Hydrated Conditions	500	8.3	2.1×10^{-6}	9.8	2.24×10^{-5}	10.9	1.07×10^{-5}	11.3	6.52×10^{-7}

Since the reinforcing fibers are made of a polymeric material, it is justified to be concern about the long-term behavior of the thermally locked GCL under sustained constant-load conditions. An attempt to understand the long-term creep shear strength behavior of the thermally locked GCL is addressed through the creep shear test. The results of the test are presented in terms of time-displacement and displacement-rate curves. Shear displacements and incremental shear rates at select times were calculated and tabulated. The test results indicate that: (i) the thermally locked GCL underwent a shear distortion immediately after application of the shear load and then started to creep and (ii) incremental shear rates continued to decrease throughout shearing for each loading phase.

The questions still remaining to be answered are: (i) how is the total shear displacement or incremental shear rate related to internal stability of the thermally locked GCL and (ii) should the total shear displacement or incremental shear rate or both be used to judge whether a thermally locked GCL is acceptable in terms of the long-term shear strength? It appears to the authors that both should be used because excessive shear deformation may affect the overall performance of the thermally locked GCL, and an increase in the incremental shear rates with time will eventually result in internal shear failure (rupture of reinforcing fibers) of the thermally locked GCL.

Although an increase in incremental shear rate indicates internal instability of the GCL, it is not quite clear whether the thermally locked GCL has fully stabilized when the incremental shear rate decreases with time. The fact is that although the incremental shear rate was decreasing with time, the duration of each time increment was not long enough to say that it had essentially reached zero. From the results of the creep shear test conducted in this research program, the incremental shear rate appears to have a linear relationship with time in the double logarithmic coordinates as shown in Fig. 9. Based on laboratory observations, it is concluded that the thermally locked GCL appears to be stable within the boundary of these test conditions since the incremental shear rate was decreasing with time.

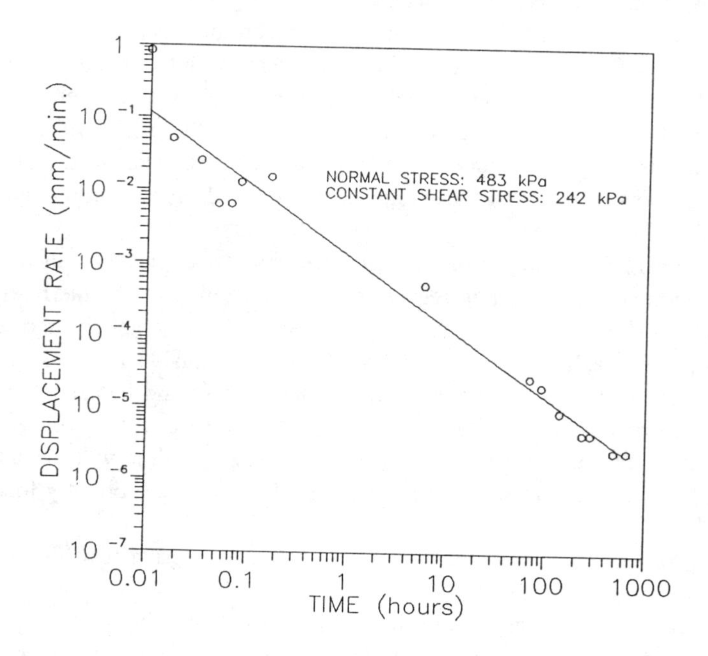

FIG. 9-- Plot of log displacement rate versus log time for the thermally locked GCL.

REFERENCES

[1] Swan, Jr., R. H., Yuan, Z., and Bachus, R. C., "**Factors Influencing Laboratory Measurement of the Internal and Interface Shear Strength of GCLS**," Testing and Acceptance Criteria for Geosynthetic Clay Liners, ASTM STP 1308, Larry W. Well, Ed., American Society for Testing and Materials, Philadelphia, 1996.

Robert J. Trauger[1], Robert H. Swan, Jr.[2], and Zehong Yuan[2]

LONG-TERM SHEAR STRENGTH BEHAVIOR OF A NEEDLEPUNCHED GEOSYNTHETIC CLAY LINER

REFERENCE: Trauger, R. J., Swan, Jr., R. H., and Yuan, Z., **"Long-term Shear Strength Behavior of a Needlepunched Geosynthetic Clay Liner,"** *Testing and Acceptance Criteria for Geosynthetic Clay Liners, ASTM STP 1308,* Larry W. Well, Ed., American Society for Testing and Materials, 1997.

ABSTRACT: This paper describes two large-scale constant-load (creep) shear testing devices that were developed to evaluate the long-term shearing behavior of geosynthetic clay liners (GCLs) and interfaces between GCLs and other geosynthetics or soils. One device was designed to simulate loading conditions that typically occur on a GCL deployed in a landfill cover system. The other device was designed to simulate loading conditions that typically occur on a GCL deployed in a landfill lining system. A needlepunched GCL was selected for evaluation of its long-term shearing behavior under these two types of loading conditions and the test results are presented in terms of time-displacement curves and shear rate-displacement curves. The results to date show that the GCL has undergone relatively small shear displacements and that the shear displacement rates within the GCL and/or at the test interface have been continuously decreasing with time. For the conditions used in this testing program, it is believed that the GCL's behavior can be considered stable. Further testing is planned to more accurately define the time-dependent internal and interface shear behavior of the GCL.

KEYWORDS: geosynthetic clay liner, shear strength, creep, direct shear, needle-punched, reinforced, landfill cover, landfill liner, constant-load creep shear test .

---

[1]Technical Services Manager, Colloid Environmental Technologies Company (CETCO), 1350 W. Shure Dr., Arlington Heights, IL 60004.

[2]Laboratory Manager and Assistant Program Manager, respectively, GeoSyntec Consultants, Soil-Geosynthetic Interaction Testing Laboratory, 5775 Peachtree Dunwoody Rd., Suite 11D, Atlanta, GA 30342.

Geosynthetic clay liners (GCLs) have been generally defined as manufactured hydraulic barriers consisting of clay bonded to a layer or layers of geosynthetic material(s). The geosynthetic components are typically geotextiles and/or geomembranes, and the clay component is usually a sodium bentonite. Commercially available GCLs consist of bentonite that is either sandwiched between two geotextiles or is bonded to a single geomembrane. GCLs may further be categorized as *unreinforced* and *reinforced*. Unreinforced GCLs have no internal reinforcement and typically possess very low shear strength. Reinforced GCLs, by means of needling, stitching, or adhesives, are designed to carry and transmit shear loads within their structure and are typically used in landfill lining and closure systems built on steep slopes. The GCL used in this testing program was a reinforced GCL, known as Bentomat ST, manufactured by Colloid Environmental Technologies Company (CETCO), Arlington Heights, Illinois, and consisted of a woven geotextile on one side of the bentonite component and a nonwoven geotextile on the other side of the bentonite component, reinforced with needlepunched fibers.

GCLs are very effective hydraulic barriers due to the flat shape of bentonite platelets and their unique ability to absorb large amounts of water. These same properties, however, cause a hydrated layer of bentonite to possess very low shear strength. Prior research [1, 2, 3] has shown that the shear strength of montmorillonite, the predominant component of bentonite, ranges from 4 to 10 degrees, depending on the species of montmorillonite and the range of applied normal stresses. Other research performed on unreinforced GCLs [4] indicates internal friction angles of 6 to 8 degrees under hydrated conditions. The internal shear strengths of reinforced GCLs are typically greater then those of unreinforced GCLs. For example the results of constant-displacement rate direct shear tests performed on a hydrated needlepunched GCL indicate a peak internal friction angle of 59 degrees under low normal stress conditions (i.e., 2 to 25 kPa) and a peak friction angle of 24 degrees under high normal stress conditions (i.e., 95 to 980 kPa) [5, 6].

While the short-term peak shear strength of a needlepunched reinforced GCL is quite high, there is growing concern that a significant portion of the short-term shear strength may be lost in the long term due to creep deterioration of the geosynthetic components and/or reinforcing materials. It is important to define "creep" within the context of this study. This term has usually been applied to single-component geosynthetics such as geotextiles, geomembranes, and geogrids where creep involves the elongation/stretching of the polymeric molecules over time. In other words, a dimensional change occurs as a result of gradual molecular reorganization with time. For the GCL used in this test program, the dimensional change is known as *displacement*, which is the lateral separation of the geotextiles in response to the applied shear force. The observed displacement may be attributable to simple elongation of geotextiles and initial stretching of needlepunched fibers, true creep of the needlepunched fibers, gradual "pullout" of the fibers, or any combination of these phenomena. Loss of the short-term shear strength of the GCL must be evaluated via adequate laboratory tests, such as large scale constant-load creep shear testing developed by the authors of this paper. Specific details of the constant-load creep shear test devices and testing procedures will be described in this paper. The constant-load shear devices were used to evaluate the long-term shear behavior of the GCL under loading conditions which typically occur in landfill cover and liner systems. The results of the tests are presented in terms of time-displacement curves and shear rate-displacement curves.

TESTING EQUIPMENT AND CALIBRATION

Equipment

Two large-scale creep shear test devices were designed and fabricated by the authors in general accordance with the equipment descriptions provided in ASTM Test Method for Determining the Coefficient of Soil and Geosynthetic or Geosynthetic and Geosynthetic Friction by the Direct Shear Method (D 5321) for constant-rate of shear testing. One device was designed to evaluate the long-term shear behavior of geosynthetics materials under low normal and shear stress conditions typically found in a landfill cover system. The other device was designed to evaluate the long-term shear behavior of geosynthetic materials under high normal and shear stress conditions typically found in a landfill liner system. The two test devices were similar in structure except for their loading systems. Each of the two test devices consisted of four major components:

- a rigid supporting table;
- an upper shear box measuring 300 mm by 300 mm in plan and 75 mm in depth and a lower shear box measuring 300 mm by 350 mm in plan and 75 mm in depth;
- a containment box to allow the test specimen to be tested under fully submerged conditions; and
- normal and shear loading systems.

For the low-pressure test device, both normal and shear loads were applied to the test specimens using dead weight. A mechanical advantage loading system was used to apply the shear force to the test specimen. The low pressure test device was designed to operate in a normal stress range of 2 to 70 kPa for a 300 mm by 300 mm test specimen. For the high-pressure test device, normal loads were applied to the test specimen through an air bladder system and shear loads through a direct loading system consisting of three 150 mm diameter air cylinders connected in parallel. The high-pressure device was designed to operate in a normal stress range of 70 to 850 kPa for a 300 mm by 300 mm test specimen. The two constant-load shear devices are schematically shown in Figs. 1 and 2.

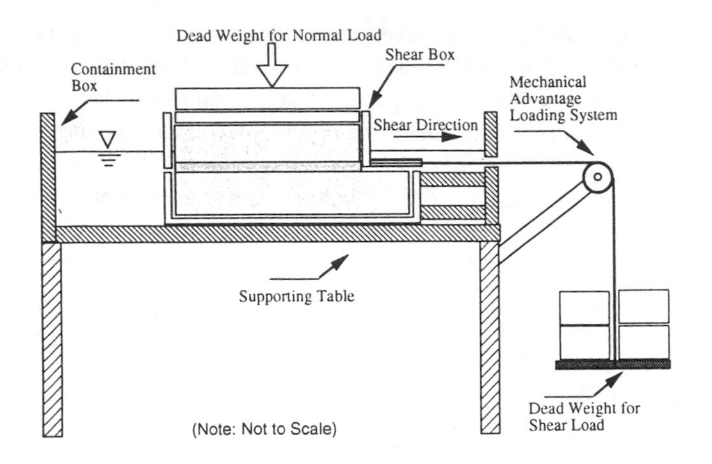

FIG. 1--Schematic of low pressure creep shear test device.

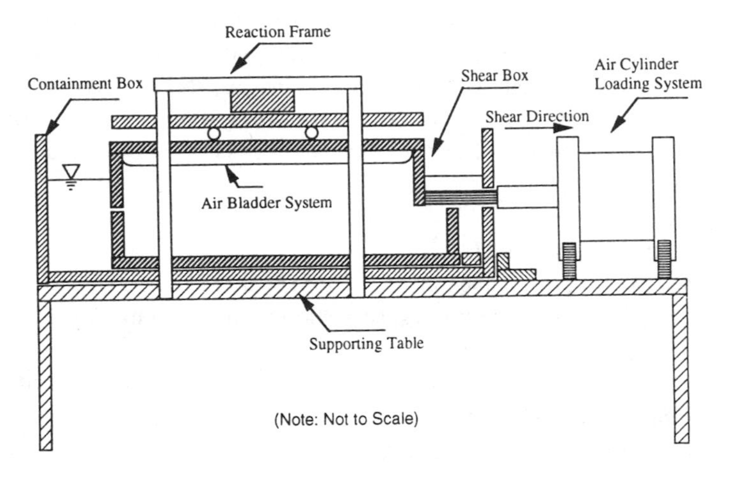

FIG. 2--Schematic of high pressure creep shear test device.

Calibration of Loading Systems

Prior to testing, calibration tests were performed on the high pressure test device to establish the relationship (i.e., calibration curve) between applied source pressures (i.e., pressure within the air cylinders and air bladder) in each loading system and actual loads delivered onto the test specimen by each loading system. Source pressures and actual normal or shear loads were measured by variable reluctance pressure transducers and electronic load cells, respectively, each recorded by a computer data acquisition system.

For the calibration of the air bladder system, an electronic load cell was placed between the air bladder system and the rigid steel reaction frame (see Fig. 2). A normal load versus pressure curve was established for source pressures ranging from 70 to 1,050 kPa. The maximum load delivered by the air bladder was approximately 80 kN at a source pressure of 1,050 kPa. It is noted that the actual normal load on top of the air bladder system is not the same as that transmitted to the test specimen. The actual applied normal load to the specimen equals the sum of the measured reaction load, weight of the air bladder system, and the test specimen itself.

For the calibration of the shear loading system (three parallel-connected air cylinders), an electronic load cell was mounted onto a steel block which was fixed to the supporting table. The load cell was then connected to the loading harness of the air cylinders through a steel rod. A shear load versus pressure curve was established for source pressures ranging from 7 to 1,050 kPa. Since the loading system was direct (no mechanical advantage), it was assumed that the applied force would remain constant and independent of displacement.

For the low pressure test device, the loads (i.e., normal and shear) were applied through dead weight, very little calibration was required. However, since a mechanical advantage loading system was used to apply the shear load, a similar calibration verification of the shear loading system was conducted except with the use of dead weight.

Trial Tests

To further confirm the two test devices were functioning properly, trial tests were conducted on an interface between a nonwoven geotextile and 1.5-mm thick smooth high density polyethylene (HDPE) geomembrane. The tests were conducted in accordance with the following test procedures:

- a fresh specimen of concrete sand was compacted into the lower shear box by hand tamping to a relatively dense state under dry conditions, forming a 75-mm thick bedding layer;
- fresh specimens of the geotextile and geomembrane were attached to the upper and lower shear boxes, respectively, with mechanical compression clamps;
- a rigid substrate was then placed on top of the geotextile specimen;
- normal loads were then applied to the test specimen through the air bladder system or using dead weight; and
- shear loads were then applied to the test specimen by increasing dead weight or air pressure within the air cylinders until sliding occurred at the test interface. The ultimate dead weight or maximum pressure within the air cylinders was recorded.

The results of the trial tests indicated that the friction angle of the geotextile-geomembrane interface was 13 degrees for normal stresses ranging from 7 to 35 kPa as measured in the low pressure test device, and 12 degrees for normal stresses ranging from 70 to 700 kPa as measured in the high pressure test device. These results are consistent with those measured by using a constant rate of shear displacement direct shear testing device (ASTM D 5321). Based on the results of these trial tests it was determined that the test devices were functioning properly.

TEST PROGRAMS

Test Materials

A needlepunched reinforced GCL (Bentomat ST) and a 2-mm thick blown-film textured HDPE geomembrane were used in the testing program. The reinforced GCL consisted of a bentonite layer sandwiched between a woven and nonwoven geotextile and reinforced with needlepunched fibers. The reinforced GCL had a typical bentonite mass per unit area of 5,000 g/m^2 and an initial (as-manufactured) moisture content of approximately 20 percent. The initial thickness of the reinforced GCL was approximately 6 mm.

Test Specimen Configurations

Two constant-load shear tests were conducted to evaluate the long-term shearing behavior of: (i) the reinforced GCL under loading conditions similar to those found in a landfill cover system and (ii) the reinforced GCL and an interface between the reinforced GCL and the textured HDPE geomembrane under loading conditions similar to those found in a landfill lining system. The configurations of the test specimens used in the two tests were as follows:

- *Test Number 1:* internal strength of the reinforced GCL under soaked and consolidated conditions. From top to bottom, the test specimen consisted of:
  - rigid substrate with textured steel gripping surface;
  - reinforced GCL;
  - rigid substrate with textured steel gripping surface; and
  - bedding layer of concrete sand.

- *Test Number 2:* internal strength of the reinforced GCL and interface between the woven geotextile of the reinforced GCL and the 2-mm textured HDPE geomembrane. From top to bottom, the test specimen consisted of:
  - rigid substrate with textured steel gripping surface;
  - reinforced GCL with woven geotextile against geomembrane;
  - 2-mm thick textured HDPE geomembrane; and
  - bedding layer of concrete sand.

It is noted that the textured steel gripping surfaces were developed by the authors during the course of establishing the testing procedures for measuring short-term internal shear strength of GCLs [5, 6]. The textured steel gripping surfaces are employed to minimize slippage between the geotextile component of the reinforced GCL and the rigid wooden substrate, therefore providing a relatively uniform transfer of shear load into the GCL specimen and/or onto the test interface.

Test Procedures and Conditions

For Test 1 (low pressure test), a fresh reinforced GCL specimen was trimmed from the bulk sample of the reinforced GCL and constrained between two rigid wooden substrates with the use of textured steel gripping surfaces. The ends of each geotextile were then sandwiched between a second rigid wooden substrate prior to testing as shown in Fig. 3. The entire test specimen was then placed in the shear box to provide confinement for the exposed bentonite component. The test specimen was then subjected to the following soaking and shearing target conditions:

1. Soaking. In tap water for 120 hours under a normal stress of 24 kPa. The normal stress used for soaking was applied prior to immersion.

2. Shearing. After the 120-hour soaking period, a constant shear load of 1.1 kN was applied to the reinforced GCL specimen using dead weight without any disruption of the soaking normal stress. The total shear load was applied to the test specimen over a period of 10 to 15 seconds in a controlled manner.

Two dial gages were used to measure vertical displacements of the GCL during soaking and shearing. Shear displacement of the GCL was measured by a dial gage attached to the upper shear box. Both vertical and shear displacements were monitored on a regular basis.

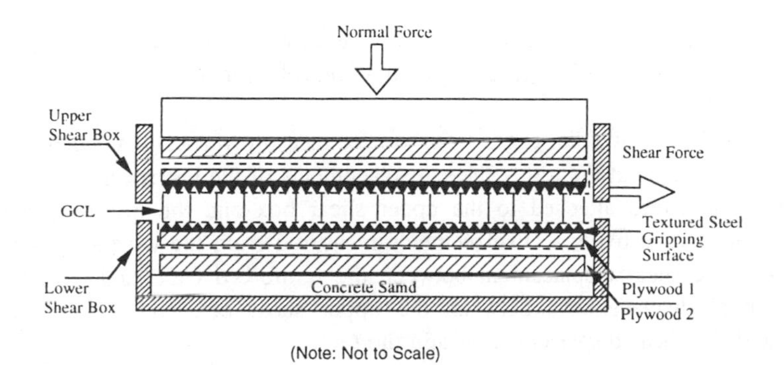

FIG. 3--Schematic of GCL test specimen configuration for Test 1.

For Test 2 (high pressure test), the concrete sand was compacted into the lower shear box by hand tamping to a relatively dense state under dry conditions, forming a 75-mm thick layer. A fresh geomembrane specimen was then placed on top of the compacted concrete sand and attached to the lower shear box with mechanical compression clamps. A fresh reinforced GCL specimen was then placed on top of the geomembrane specimen with its woven geotextile component in contact with the geomembrane. A rigid substrate with a textured steel gripping surface was placed on top of the reinforced GCL. The complete configuration of Test 2 is shown in Fig. 4. The test specimen was then subjected to the following soaking, consolidation, and shearing target conditions:

1. Soaking. In tap water for 120 hours under a normal stress of 19 kPa which was applied using dead weight. The normal stress used for soaking was applied prior to immersion.

2. Consolidation and Shearing Phase 1. Consolidated for 120 hours under a normal stress of 97 kPa and then subjected to a constant shear stress of 34 kPa for 1,000 hours.

3. Consolidation and Shearing Phase 2. Consolidated for 120 hours under a normal stress of 194 kPa and then subjected to a constant shear stress of 68 kPa for 1,000 hours.

4. Consolidation and Shearing Phase 3. Consolidated for 120 hours under a normal stress of 292 kPa and then subjected to a constant shear stress of 102 kPa for 1,000 hours.

5. Consolidation and Shearing Phase 4. Consolidated for 120 hours under a normal stress of 389 kPa and then subjected to a constant shear stress of 136 kPa for 1,000 hours.

6. For each loading phase, the shear load was applied to the test specimen over a period of 10 to 15 seconds in a controlled manner.

Similar to Test 1, two dial gages were used to measure vertical displacements of the test specimen throughout the test. Shear displacements of the test specimen were measured by two dial gages attached to the upper shear box and the lower woven geotextile component, respectively. Displacements measured from the woven geotextile directly indicated the shear displacement between the reinforced GCL and the geomembrane. Differential displacements between the upper shear box and the woven geotextile indicated the shear displacement within the GCL.

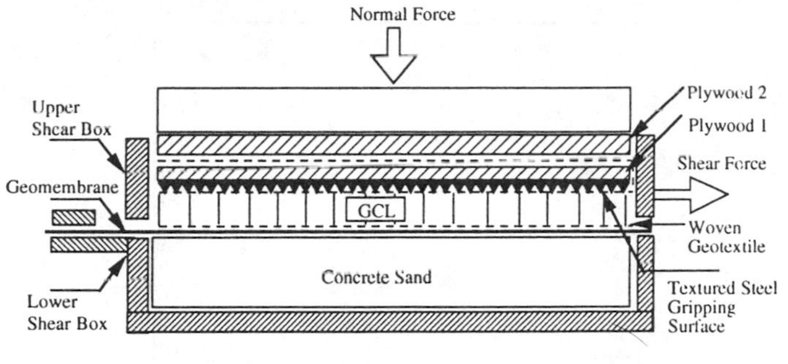

FIG. 4--Schematic of GCL test specimen configuration for Test 2.

The ratio of shear load to normal load used in Tests 1 and 2 were 1:2 and 1:5 respectively. Test 1 simulated loading conditions on a GCL used in a landfill closure system constructed on a 2H:1V side slope. Test 2 simulated loading conditions on a GCL used in a landfill liner system constructed on a 5H:1V side slope. These loading conditions imposed onto each test specimen are indicated on the short-term shear strength plots for the two test specimens (Figs. 5 and 6). The constant shear stress was approximately 23 percent of the peak short-term shear strength of the reinforced GCL for Test 1 as indicated in Fig. 5. For Test 2, the constant shear stress varied from approximately 40 to 70 percent of the peak short-term shear strength of the reinforced GCL as indicted in Fig. 6. It is also noted that the short-term interface shear strength of a woven geotextile component of a reinforced GCL against a textured geomembrane is typically lower than that of the internal shear strength of the reinforced GCL.

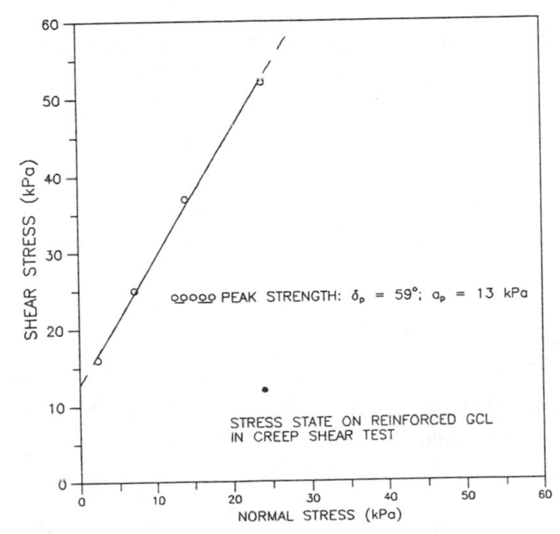

FIG. 5-- Plot of the loading condition for Test 1 on the reinforced GCL with the short-term peak shear strength envelope of the GCL at low normal stresses [5].

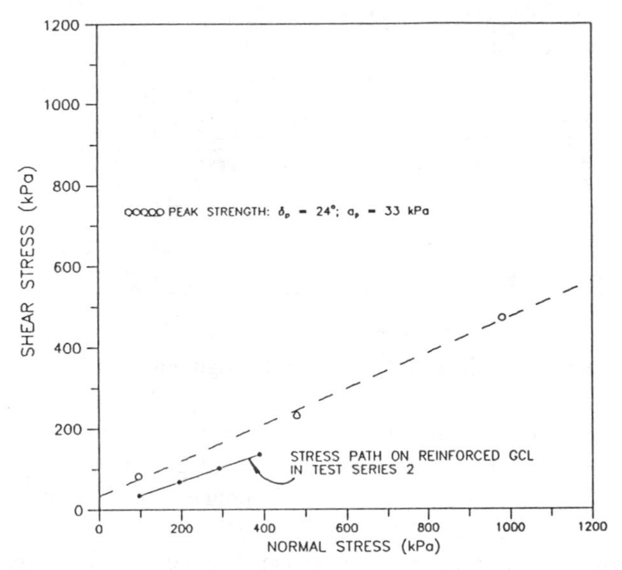

FIG. 6-- Plot of the loading path for Test 2 on the reinforced GCL-geomembrane interface with the short-term peak shear strength envelope of the GCL at high normal stresses [6].

TEST RESULTS

Test Series 1 (Low Pressure Test)

The results of the constant-load creep shear test (Test 1) conducted on the reinforced GCL are presented graphically in Figs. 7 through 10. Fig. 7 indicates that the reinforced GCL was initially compressed and then swelled under the applied soaking normal stress. Fig. 8 indicates that the reinforced GCL specimen underwent a slight shear distortion immediately after application of the shear load and that it compressed slightly during shearing. Figs. 9 and 10 indicate incremental shear displacement rates decreased from approximately 12 mm/min at the beginning of shearing to approximately 1 x 10^{-7} mm/min.

A summary of shear displacement and incremental shear rates at select times is presented in Table 1. This table presents the total shear displacements and incremental shear rates at 1, 10, 100, 500, 1,000, and 10,000 hours of elapsed time for the creep shear test. It is noted that the values of total displacement and incremental shear rate at these selected times presented in Table 1 were derived by linear interpolation between the actually measured data points at each specific time period of interest.

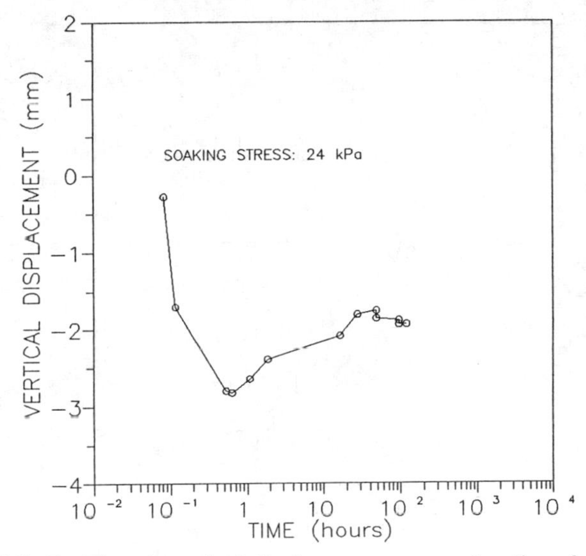

FIG. 7-- Plot of vertical displacement versus log time for Test 1.

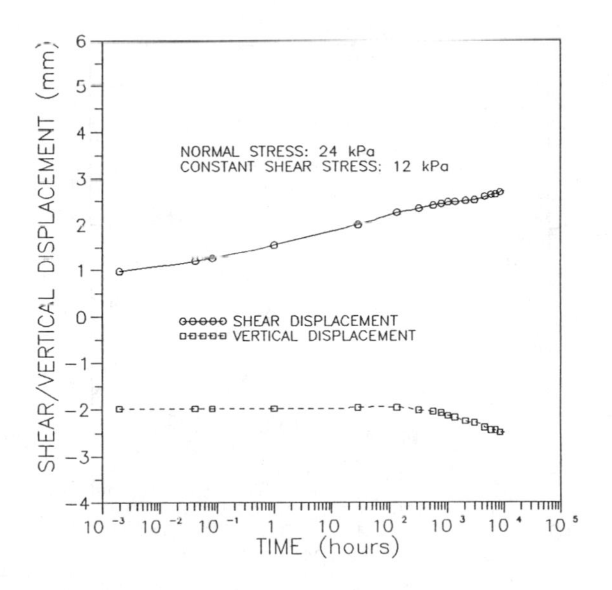

FIG. 8-- Plot of shear and vertical displacements versus log time for Test 1.

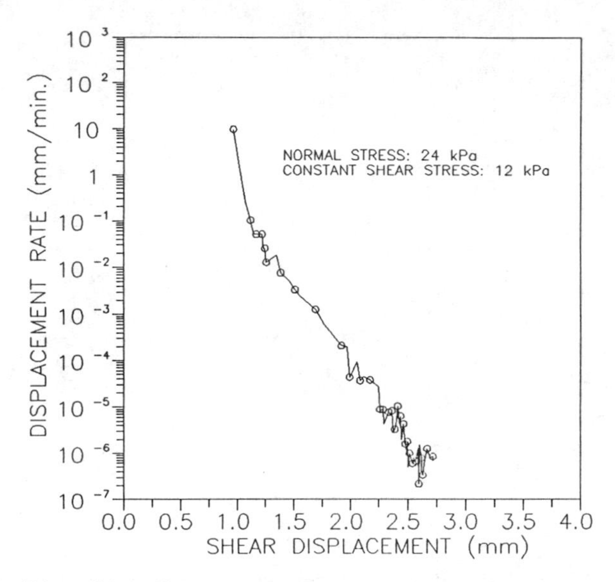

FIG. 9-- Plot of log displacement rate versus shear displacement for Test 1.

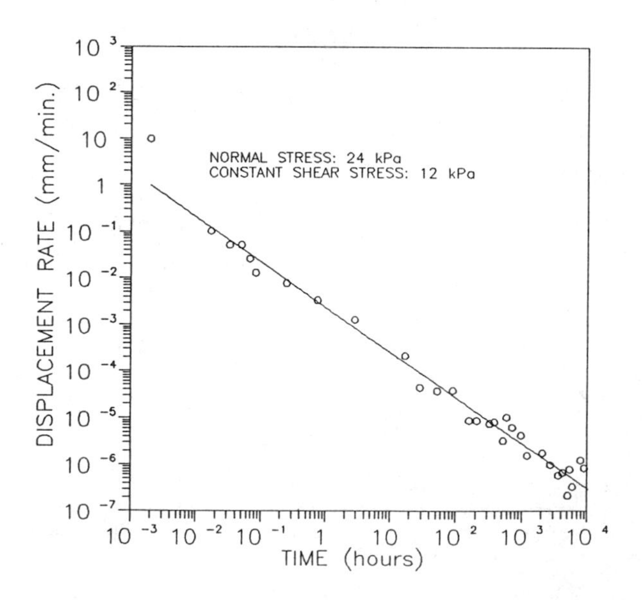

FIG. 10-- Plot of log displacement rate versus log time for Test 1.

TABLE 1--Summary of Test 1 creep shear test results.

Test Number 1 Test Specimen and Loading Conditions	Elapsed Time (hours)	Total Shear Displacement (mm)	Incremental Shear Rate (mm/min)
Needle-punched GCL Under Hydrated Conditions	1	1.5	3.38×10^{-3}
	10	1.7	9.93×10^{-4}
	100	2.1	4.22×10^{-5}
	500	2.4	4.72×10^{-6}
Normal Load: 1.1 kN	1,000	2.5	2.84×10^{-6}
Shear Load: 0.56 kN	10,000	2.7	5.49×10^{-7}

Test Series 2 (High Pressure Test)

The reinforced GCL and reinforced GCL-geomembrane interface in Test 2 were subjected to an initial soaking followed by four consolidation/shearing phases. Compression and swelling of the test specimen during the soaking phase are shown in Fig. 11. The results of a typical loading phase are graphically presented in Figs. 12 through 15 in terms of vertical/shear displacements versus logarithm of time, logarithm of shear displacement rate versus shear displacement, and logarithm of shear displacement rate versus logarithm of time.

A summary of shear displacement and incremental shear rates at select times is presented in Table 2. This table presents the total shear displacements and incremental shear rates at 1, 10, 100, 500, and 1,000 hours of elapsed time for each phase of the creep shear test. It is noted that the values of total displacement and incremental shear rate at these selected times presented in Table 2 were derived by linear interpolation between the actually measured data points at each specific time period of interest. It should be also noted that since this was a research study, some of the loading phases were maintained and monitored longer than the targeted 1000 hours (i.e., up to 1900 hours for one of the phases). The behavior of these extended loading phases was found to follow the trends established in the first 1000 hours. For comparison purposes the data for the first 1000 hours of each loading phase are presented. The overall test duration of Test 2 was approximately 7,200 hours, including all of the soaking, consolidation, and shearing phases.

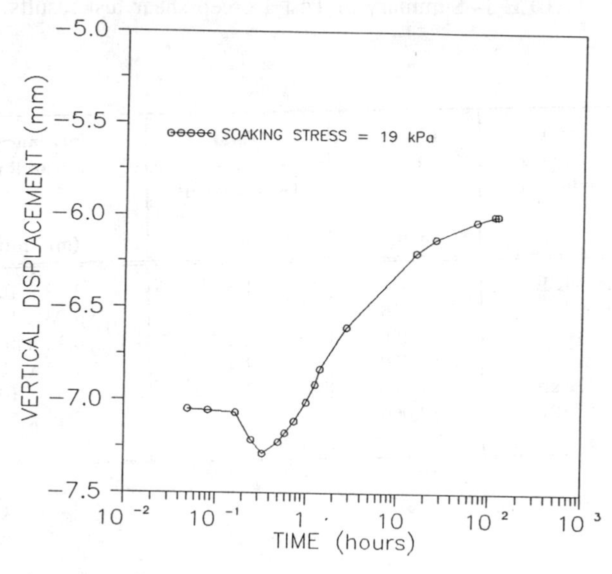

FIG. 11-- Plot of vertical displacement versus log time for the soaking phase of Test 2.

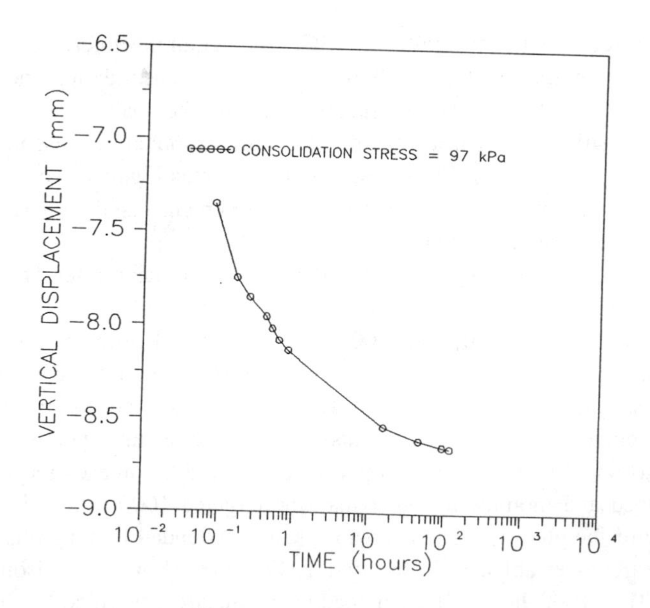

FIG. 12-- Plot of vertical displacement versus log time for one of the loading phases of Test 2.

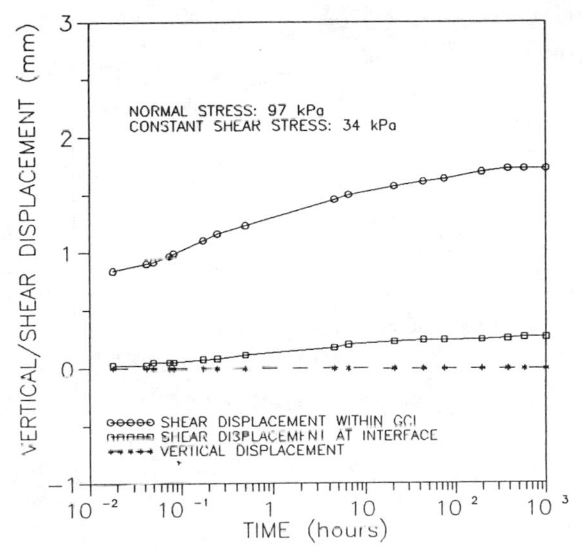

FIG. 13-- Plot of shear and vertical displacements versus log time for one of the loading phases of Test 2.

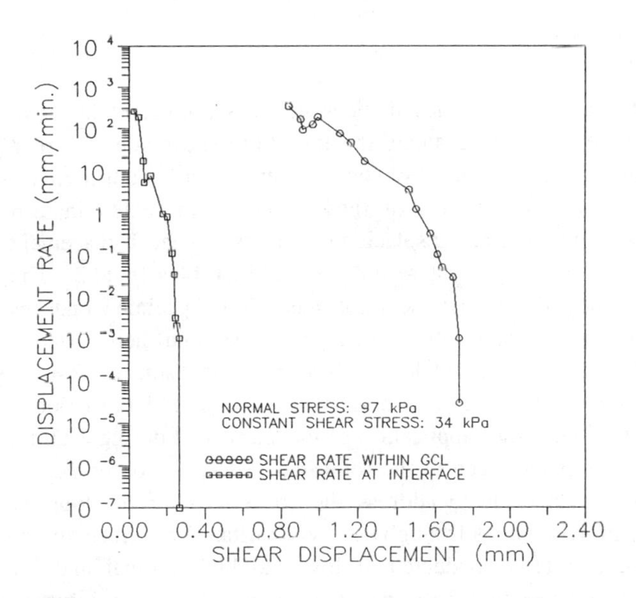

FIG. 14-- Plot of log displacement rate versus shear displacement for one of the loading phases of Test 2.

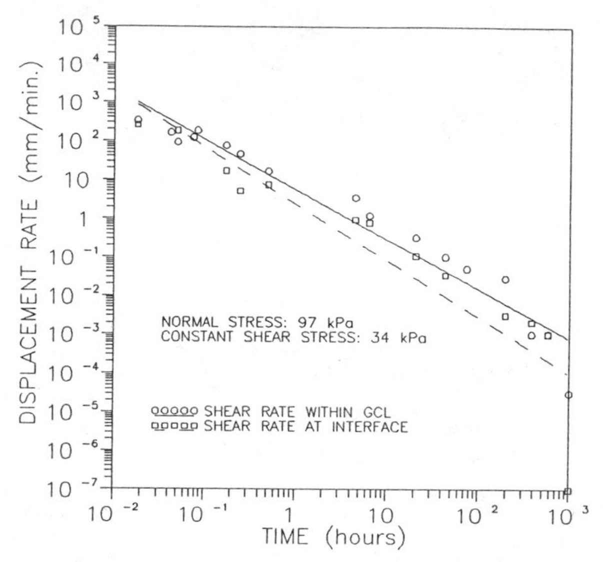

FIG. 15-- Plot of log displacement rate versus log time for one of the loading phases of Test 2.

SUMMARY AND DISCUSSIONS

Two constant-load creep shear test devices were designed and fabricated. The creep shear test devices were used to evaluate the long-term shear behavior of a reinforced GCL under loading conditions typical of those occurring in a landfill cover system and a landfill liner system. The results of these tests are presented in terms of time-displacement curves and shear rate-displacement curves. Shear displacements and shear rates for the two tests at select times are summarized in Tables 1 and 2. The test results of the two tests indicated that: (i) the reinforced GCL specimen underwent a shear distortion immediately after application of the shear load and then started to creep and (ii) incremental shear rates decreased during shearing. The results of Test 2 also indicted that the reinforced GCL specimen underwent a small "translation" along the geomembrane immediately after application of the shear load during each loading phase and then seemed to stop and creep of the reinforced GCL was observed.

Attempts have been made to address the concerns over the long-term strength behavior of the reinforced GCLs through the two constant-load creep shear tests. Data obtained from the two tests conducted at low and high normal and shear stresses indicated that the total accumulated shear displacement within the reinforced GCL or at the test interface was on the order of 0.4 mm (Test 2 interface) to 2.7 mm (Test 1 GCL) or 3.5 mm (Test 2 GCL). It can also be seen from the data for both tests that the

TABLE 2--Summary of Test 2 creep shear test results.

Test Specimen and Loading Conditions	Elapsed Time (hours)	Total Shear Displacement of GCL (mm)	Total Shear Displacement of Interface (mm)	Incremental Shear Rate of GCL (mm/min.)	Incremental Shear Rate of Interface (mm/min.)
Needle-Punched GCL/80-mil Textured HDPE Geomembrane Under Soaked Conditions	1	1.3	0.1	1.47×10^{-1}	6.6×10^{0}
	10	1.5	0.2	9.78×10^{-1}	6.27×10^{-1}
	100	1.7	0.3	4.47×10^{-2}	6.25×10^{-4}
Normal Load: 8.9 kN	500	1.7	0.3	4.17×10^{-3}	3.15×10^{-4}
Shear Load: 3.2 kN	1,000	1.7	0.3	3.12×10^{-5}	1.06×10^{-7}
Needle-Punched GCL/80-mil Textured HDPE Geomembrane Under Soaked Conditions	1	2.1	0.3	1.57×10^{-3}	0.00×10^{-0}
	10	2.2	0.3	8.79×10^{-4}	0.00×10^{-0}
	100	2.3	0.3	2.36×10^{-5}	0.00×10^{-0}
Normal Load: 17.8 kN	500	2.4	0.3	4.78×10^{-6}	0.00×10^{-0}
Shear Load: 6.2 kN	1,000	2.5	0.3	4.32×10^{-6}	0.00×10^{-0}
Needle-Punched GCL/80-mil Textured HDPE Geomembrane Under Soaked Conditions	1	2.6	0.3	4.14×10^{-4}	0.00×10^{-0}
	10	2.6	0.3	1.94×10^{-4}	0.00×10^{-0}
	100	2.7	0.3	2.67×10^{-5}	0.00×10^{-0}
Normal Load: 26.7 kN	500	2.8	0.3	2.62×10^{-6}	0.00×10^{-0}
Shear Load: 9.4 kN	1,000	2.8	0.3	1.74×10^{-6}	0.00×10^{-0}
Needle-Punched GCL/80-mil Textured HDPE Geomembrane Under Soaked Conditions	1	3.0	0.4	1.32×10^{-4}	0.00×10^{-0}
	10	3.1	0.4	1.71×10^{-4}	0.00×10^{-0}
	100	3.3	0.4	9.27×10^{-6}	0.00×10^{-0}
Normal Load: 35.6 kN	500	3.5	0.4	1.26×10^{-7}	0.00×10^{-0}
Shear Load: 12.5 kN	1,000	3.5	0.4	1.11×10^{-7}	0.00×10^{-0}

majority of the total accumulated shear displacement happened within the first 100 hours of each test/loading phase. Hence the actual time-dependent displacement (creep) is probably on the order of 0.2 to 0.6 mm for each test/loading phase. For the reinforced GCL in both tests, incremental shear rates decreased with time from approximately 1×10^{-1} mm/min to approximately 1×10^{-7} mm/min during shearing. For the reinforced GCL-geomembrane interface, incremental shear rates decreased rapidly to zero after application of the shear load.

While no other long-term shear test data exist at this time, these results are consistent with previous project experience of the authors and the large-scale field test in progress in Cincinnati, USA, [7]. These results are also consistent with short-term testing performed on the reinforced GCL which indicates that the reinforced GCL is capable of sustaining greater loads than were applied during each of the long-term tests. More testing will be performed to further assess the long-term internal and interface shear strength behavior of the reinforced GCL under various loading conditions.

REFERENCES

[1] Olson, R. E., "**Shearing Strength of Kaolinite, Illite, and Montmorillonite**," Journal of Geotechnical Engineering, ASCE No. GT11, pp.1215-1227, 1974.

[2] Mesri, G. and Olson, R. E., "**Shear Strength of Montmorillonite**," Geotechnique, Vol. 20, No. 3, pp.261-270, 1970.

[3] Gleason, M. H., "**Comparative Testing of Calcium and Sodium Smectite Clays for Geotechnical Environmental Applications**, M.S. Thesis, University of Texas, Austin, Texas, 200 pp, 1993.

[4] Daniel, D. E. and Shan, H.-Y., "**Results of Direct Shear Tests on Hydrated Bentonitic Blankets**," University of Texas, Geotechnical Engineering Center, Austin, Texas, 13 pp, 1991.

[5] GeoSyntec Consultants, "**Final Report, Direct Shear Testing, Internal Strength of Bentomat GCL**," Project No. GL3419, 7 December 1993.

[6] GeoSyntec Consultants, "**Final Report, Direct Shear Testing, Internal Strength of Bentomat GCL Under High Normal Stresses**," Project No. GL3529, 2 February 1994.

[7] Daniel, D. E., Interim Data from Cincinnati GCL Field Test Site. Unpublished.

Robert B. Gilbert[1], Heather B. Scranton[2] and David E. Daniel[3]

SHEAR STRENGTH TESTING FOR GEOSYNTHETIC CLAY LINERS

REFERENCE: Gilbert, R. B., Scranton, H. B., and Daniel, D. E. **"Shear Strength Testing for Geosynthetic Clay Liners,"** *Testing and Acceptance Criteria for Geosynthetic Clay Liners, ASTM STP 1308,* Larry W. Well, Ed., American Society for Testing and Materials, 1997.

ABSTRACT: This paper provides guidelines for measuring the internal and interface shear strengths of geosynthetic clay liners (GCLs) in a direct shear test. Currently, there is significant variability between laboratories in shear testing procedures for GCLs; the major differences are related to specimen hydration and shear rate. The guidelines presented in this paper are intended to provide for consistent test results that can be interpreted without ambiguity. They are also intended to provide measured strengths that are representative of typical field conditions.

KEYWORDS: shear strength, internal strength, interface strength, geosynthetic clay liner, bentonite, geosynthetics, testing procedure, direct shear

---

This paper provides guidelines for using a direct shear test to measure the internal and interface shear strengths of geosynthetic clay liners (GCLs). Shear strength is an important design consideration for GCLs used in waste containment applications. Sodium bentonite has the lowest shear strength among common clay soils, and many of the interfaces with GCLs also exhibit low strengths. Laboratory determined strengths exhibit wide variability that is related to different test procedures and conditions. For example, Fig. 1 shows the scatter of different test results for the peak, internal shear strength of a hydrated, needle-punched GCL. The test guidelines presented in this paper are designed to meet the following objectives:

---

[1] Assistant Professor, Department of Civil Engineering, The University of Texas at Austin, Austin, TX 78712.
[2] Graduate Research Assistant, Department of Civil Engineering, The University of Texas at Austin, Austin, TX 78712.
[3] Professor, Department of Civil Engineering, The University of Texas at Austin, Austin, TX 78712.

1. produce test results that are consistent between different laboratories;
2. produce test results that can be interpreted without ambiguity; and
3. produce measured shear strengths that are representative for typical field conditions.

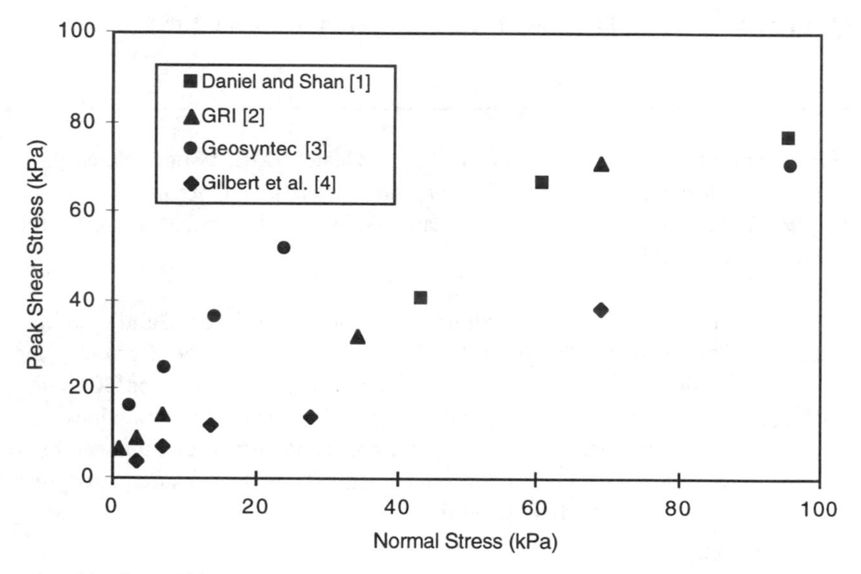

Fig. 1--Measured peak, internal shear strength for a hydrated, needle-punched GCL.

DEFINITION OF TYPICAL FIELD CONDITIONS

The shear strength of a GCL in the field depends on the hydration and loading conditions in the field. For the purpose of designing appropriate shear test guidelines that are representative of typical field conditions, typical field conditions are defined as follows:

The GCL is hydrated. The bentonite in as-produced GCLs is at a nominal moisture content of 10 to 30 percent. This moisture content corresponds to a matric suction that is extremely high relative to most natural soils; therefore, the bentonite will tend to draw water from adjacent soil layers and become hydrated. For example, Daniel et al. [5] demonstrate that a GCL will generally hydrate to moisture contents in excess of 100 percent even when it is in contact with very dry soils having moisture contents less than 10 percent. GCLs in liners can also become hydrated with the waste liquids they are intended to contain, while GCLs in covers can become hydrated from precipitation and moisture in landfill gases. A possible exception to this typical field condition is a GCL that is encased between two geomembranes and installed in an arid environment.

The GCL is subjected to static normal and shear stresses. Most GCLs in liners and covers are placed on or near slopes and beneath fill (Fig. 2). The weight of fill over the GCL applies normal and shear stresses to the GCL. Clearly, these stresses will increase during construction or waste placement, and are not strictly static. However, they can be considered static if they are applied slowly enough for the GCL to maintain equilibrium (i.e., for the bentonite to consolidate and drain in response to the increasing stress). The loading rate producing drained conditions can be estimated from Terzaghi's theory of one-dimensional consolidation for an applied stress that increases linearly with time [6]. Table 1 presents the estimated time required for 95-percent consolidation of a GCL under different final normal stresses. The consolidation parameters used in these calculations were obtained from Shan [7], and are representative for both unreinforced and reinforced GCLs. The results in Table 1 indicate that a GCL will come to equilibrium very quickly relative to typical load application rates in the field. For example, a normal stress of 170 kPa corresponds to approximately 17 m of waste; it would be difficult to place 17 m of waste in less than 16 days. Therefore, GCLs in typical field conditions will be subjected to fully-drained loading conditions. An exception is earthquake loading.

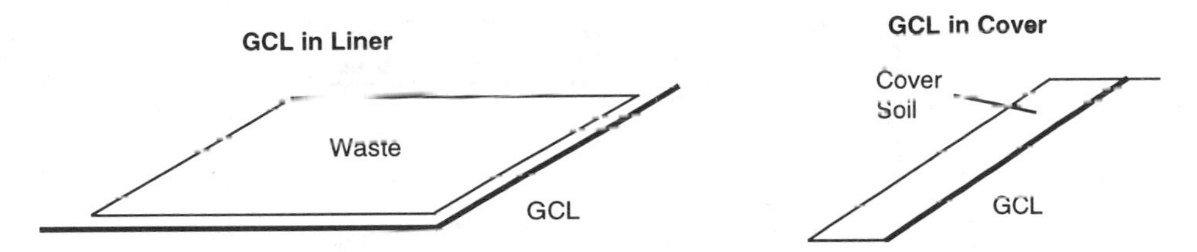

Fig. 2--Typical field loading conditions for a GCL.

Table 1--Estimated time required for 95-percent consolidation of GCL under an applied load that increases linearly with time.

	Time Required for 95-Percent Consolidation		
GCL Drainage Conditions	Final Normal Stress at End of Loading: 17 kPa	Final Normal Stress at End of Loading: 170 kPa	Final Normal Stress at End of Loading: 690 kPa
Drainage on both Sides of GCL	1.5 days	4 days	3 days
Drainage on one Side of GCL	6 days	16 days	12 days

HYDRATION

Hydration of the GCL can have a significant effect on its shear strength. First, hydration affects the shear strength of bentonite. The peak internal shear strength of an unreinforced GCL versus moisture content is shown on Fig. 3 for two different normal stresses. The shear strength decreases as the moisture content increases. Second, for GCLs with bentonite encased between geotextiles, the bentonite may extrude through the geotextiles into adjacent interfaces and affect the interface strength. Bentonite extrusion is normally associated with woven geotextiles, although it has been observed for thin (i.e., mass per unit area less than 220 g/m^2) nonwoven geotextiles as well. An example of bentonite extrusion is shown on Fig. 4. This specimen was hydrated with distilled water under a normal stress of 14 kPa. Not only did bentonite extrude through the woven geotextile, but a smooth geomembrane that was adjacent to the woven geotextile was also smeared with bentonite after hydration. Finally, hydration of bentonite may affect the properties of reinforced GCLs by stretching the reinforcement as the bentonite swells.

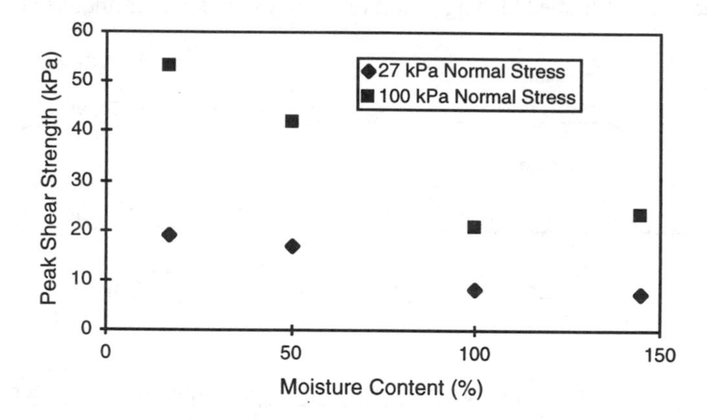

Fig. 3--Drained[1], peak internal shear strength versus moisture content for an unreinforced GCL (data from [5]).

If hydration of the GCL in the field is possible, it is important to hydrate the bentonite completely in the laboratory before shear to obtain a representative shear strength. Complete hydration before shear is also necessary to produce test results that can be interpreted without ambiguity because a partially hydrated specimen will continue to hydrate during shear. The completeness of hydration can be determined in the laboratory by measuring the thickness of the GCL as it hydrates; complete hydration is

---

[1] A shear rate of 0.0003 mm/min was used, which corresponds to drained conditions for the fully hydrated GCL under these normal stresses.

achieved when the thickness reaches a constant value. In practical terms, hydration progress can be monitored by the percentage change in thickness over a 12-hour period:

$$\text{Hydration Factor} = \frac{\left| h_t - h_{t-12\,hr} \right|}{h_t} \qquad (1)$$

where

h_t = GCL height at time t
$h_{t\text{-}12\ hr}$ = GCL height 12 hours before time t

Fig. 4--Photograph of bentonite extrusion through the woven geotextile of a needle-punched GCL that was hydrated with distilled water under a 14 kPa normal stress.

Hydration Factors of less than 5 percent generally can be obtained within reasonable time periods of ten to twenty days. The time required for complete hydration (i.e., Hydration Factor < 5 percent) versus the applied normal stress is shown on Fig. 5 for needle-punched GCL specimens that were hydrated with distilled water. The specimens were 280 mm by 430 mm in size, and drainage was provided along only one side of the GCL (the other side was against a rigid backing plate or a geomembrane). The required hydration time tended to decrease with increasing normal stress because the change in height was smaller at higher normal stresses. There was significant variability in the required hydration times, possibly indicating the magnitude of variability in the bentonite properties.

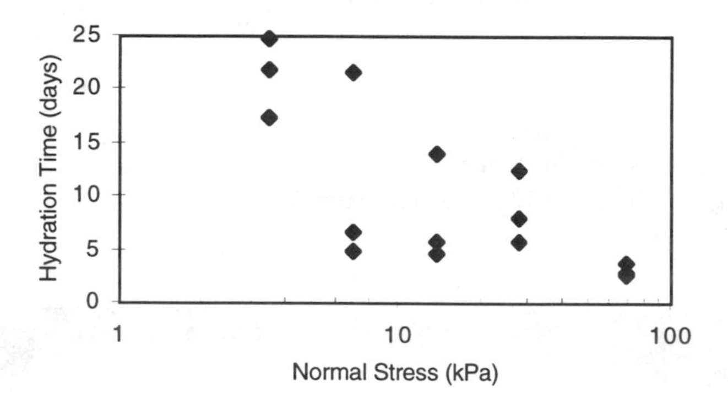

Fig. 5--Time required for complete hydration versus normal stress for a needle-punched GCL hydrated with distilled water and single drainage (data from [4]).

The order of hydration and normal stress application is an issue that requires consideration in testing. A GCL in a lining system could hydrate under a small normal stress after installation and then consolidate under larger normal stresses as the waste is placed. Conversely, a GCL in a cover could hydrate under the final normal stress (i.e., the weight of the cover). For unreinforced GCLs, the order of hydration and normal stress application should not affect the bentonite's strength appreciably. This conclusion is supported with consolidation and swelling data presented by Shan [7] for unreinforced GCL products; the final void ratio of the bentonite was not very sensitive to the order of hydration and normal stress application. Therefore, hydration under the final normal stress is recommended for unreinforced GCLs since it will require less time than initial hydration and subsequent consolidation. However, the order of hydration may affect the measured shear strength for reinforced GCLs. First, it may affect the reinforcement capacity because the reinforcement will be stretched more if the GCL is hydrated first under a small normal stress. Second, interface strengths may be affected since more bentonite may extrude into the interface if the bentonite is hydrated first under a small normal stress and then consolidated. Therefore, the order of hydration and normal stress application in direct shear testing for reinforced GCLs should model the expected conditions in the field as closely as possible.

It is important to provide free drainage along at least one side of the GCL during the hydration stage. Fig. 6 shows a photograph of a 280 mm by 430 mm specimen of a needle-punched GCL that had seemingly reached complete hydration, as determined by no further change in thickness. Rigid backing plates were located both above and below the GCL during hydration. Although the carrier geotextiles (a woven and a non-woven) allowed some water transmission into the bentonite layer, the water did not reach the interior of the specimen. The shortest hydration times will be achieved if both sides of the GCL are freely draining during hydration. However, it is very important in testing GCL interfaces that the interface material be against the GCL during hydration due to the potential for bentonite extrusion into the interface. Therefore, only one freely draining boundary can be counted on when testing interfaces between GCLs and geomembranes.

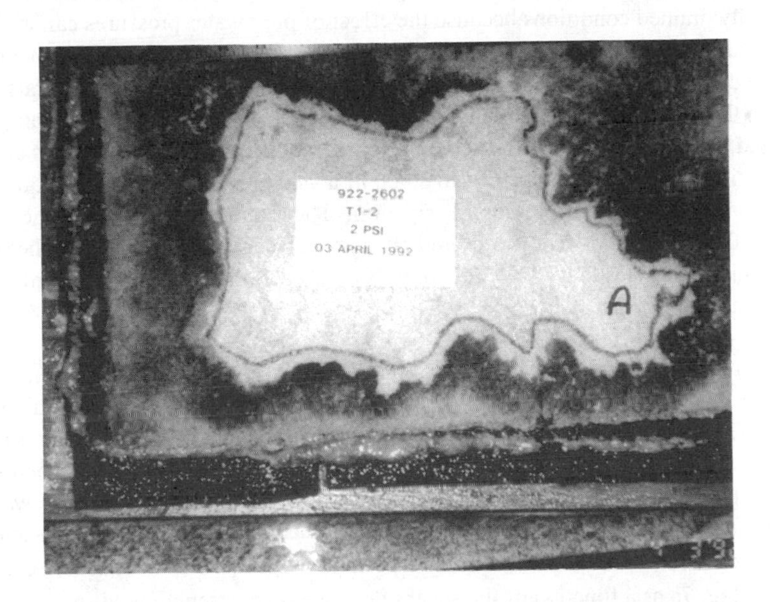

Fig. 6--Photograph of incomplete hydration in a needle-punched GCL specimen (280 mm by 430 mm in size) due to poor drainage during hydration.

The magnitude of swell in bentonite due to hydration is related to the hydration fluid. Distilled water causes the greatest amount of swelling, while inorganics and organic solvents tend to produce less swelling. The internal and interface strengths for a GCL will tend to decrease with increased swelling (i.e., greater moisture content in the bentonite, more potential for bentonite extrusion into interfaces, and more stretching of reinforcement). While the most conservative hydration fluid is therefore distilled water, it is not likely to be representative of field conditions. Tap water is recommended in the ASTM standard for hydraulic conductivity testing (ASTM D-5084) because it is readily available and its chemistry is assumed to be comparable to typical pore water in soils. Based on similar reasoning, tap water is recommended as the hydration fluid for testing the shear strength of GCLs.

RATE OF SHEAR

The rate of shear in a direct shear test is important because pore water pressures that can affect the strength may be generated within the bentonite during shear. In a direct shear test, drainage cannot be prevented and pore water pressures cannot be measured; hence, it is important to shear the specimen slowly enough so that shear-generated pore water pressures dissipate. Otherwise, there will be ambiguity in intrepreting test results for partially drained conditions because the effect of pore water pressures cannot be determined. ASTM D3080 provides guidance on the rate of shear required to achieve complete drainage of shear-generated pore water pressures in soils: the shear rate is related to the rate of consolidation in a one-dimensional consolidation test. Shan [7] conducted one-dimensional consolidation tests on four different GCL products at consolidation stresses ranging from 4 to 3,000 kPa, and then determined the required shear rate to allow for pore water pressure dissipation. The required shear rates ranged from 0.0001 to 0.001 mm/min. These required rates are substantially smaller than the 0.01 to 1.0 mm/min shear rates that have commonly been used to measure GCL strengths in commercial laboratories.

Direct shear tests were conducted on a hydrated, unreinforced GCL to investigate how shear rate affects the strength of the bentonite. The test specimens had a diameter of 63 mm. They were hydrated with tap water for 24 hours under the normal stress that was applied during shear, 17 kPa or 170 kPa. Several shear tests were conducted with hydration times of 48 and 72 hours to confirm that the 24 hour hydration time was adequate. Two to three direct shear tests were conducted at each normal stress and shear rate to account for variability. The ratio of peak strength, τ_p, to normal stress, σ_n, is shown on Fig. 7a as a function of the shear rate. The shear strength tends to decrease with decreasing shear rate, especially at the lower normal stress where the strength at a rate of 0.0005 mm/min is about 60 percent of the strength at a rate of 1.0 mm/min. This result is consistent with other published data, which are more limited in scope, for both unreinforced [1] and reinforced GCLs [4].

The reduction in strength with rate may be due to pore water pressures, or it may be related to other rate-dependent mechanisms such as creep [8]. A one-dimensional consolidation test was performed on this GCL to determine the relationship between excess pore water pressure and shear rate. For each normal stress, the coefficient of consolidation measured in the consolidation test was incorporated into the model by Gibson and Henkel [9] to relate pore water pressure drainage to time. Time was then converted to a shear rate by assuming a deformation at failure of 2 mm. The results, in terms of the ratio of excess pore water pressure to the intial pore water pressure, are shown on Fig. 7b for the two normal stresses. The strength seems to reach a constant value for shear rates near those required for complete drainage, about 0.001 mm/min.

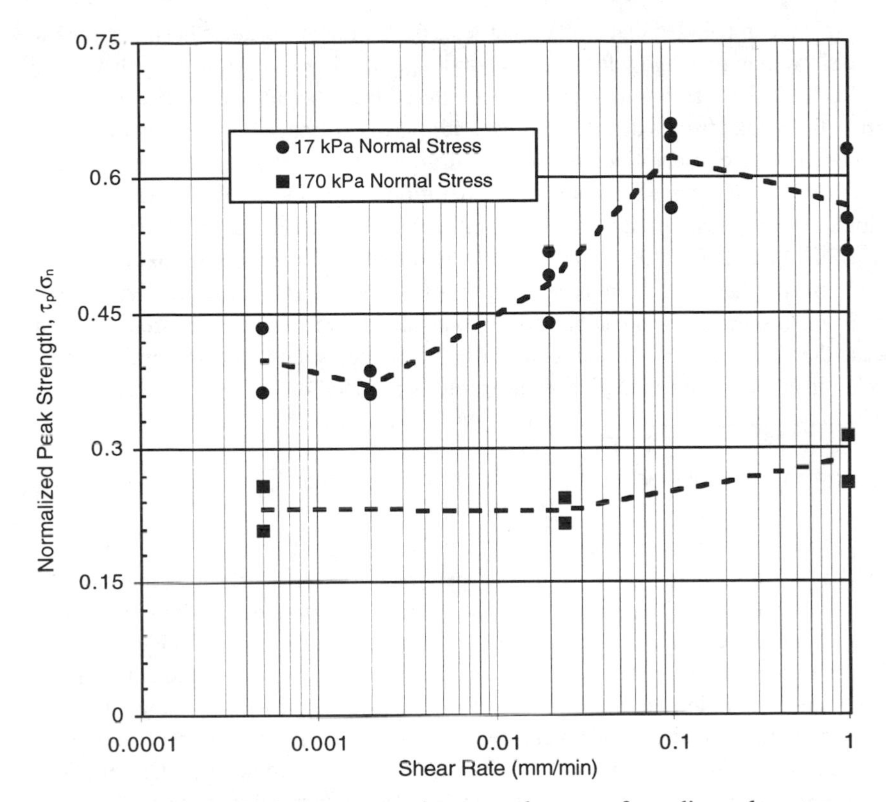

(a) Normalized peak shear strength versus shear rate from direct shear tests.

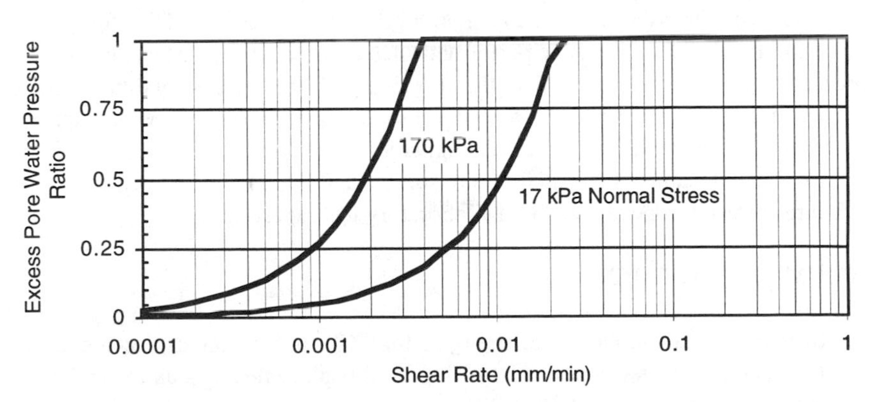

(b) Ratio of excess to initial pore water pressure versus shear rate calculated from consolidation test results.

Fig. 7--Direct shear and consolidation test results for an unreinforced GCL.

It is recommended that direct shear tests on GCLs be conducted using shear rates that are slow enough to allow for complete drainage of the bentonite, in accordance with ASTM D3080. This recommendation has a rational basis because it is difficult to interpret the results from a direct shear test with partial drainage. This recommendation also appears to be conservative. As in the hydration stage, the drainage conditions at the boundary of the GCL are an important consideration during the shearing stage. If possible, porous discs should be placed on either side of the GCL to provide for drainage. If one of the boundaries cannot drain freely (e.g., because of the presence of a geomembrane), then the required shear rate from ASTM D3080 should be divided by four. Drained shear rates should be used in interface, as well as internal, strength testing because the strength of the bentonite may affect interface strengths. Further research on the relationship between strength and shear rate is needed both for bentonite and for the geosynthetic components in GCLs and at interfaces with GCLs.

FAILURE SURFACE LOCATION

The failure surface (i.e., the plane at which slippage occurs in the specimen) should be free to form anywhere within the specimen. In internal shear testing, the failure surface can be located within the bentonite layer or at the interface between the bentonite and one of the carrier materials. For example, the internal failure plane in one reinforced GCL was located at the interface between the bentonite and the woven geotextile encasing the bentonite [4]. In interface testing, the failure surface can be located at the interface or within the GCL. Gilbert et al. [4] presented interface test results for smooth and textured geomembranes against a reinforced GCL. For the smooth geomembrane/GCL interface, the failure surface was located at the interface. For the textured geomembrane/GCL interface, the failure surface was located at the interface for normal stresses less than about 15 kPa, and within the reinforced GCL at higher normal stresses. Therefore, it is important that the direct shear test be performed such that the failure surface location is not forced, but free to develop anywhere through the specimen. If this condition cannot be achieved (e.g., the failure surface is forced to form at the location between the upper and lower box in a conventional direct shear apparatus), then multiple tests are required to measure the strength along all potential failure surfaces. It is imperative that the location of the failure surface be documented in the laboratory test report.

MAXIMUM DISPLACEMENT

The internal and interface shear strengths for GCLs depend on displacement. A peak strength is typically developed at 2 to 10 mm of displacement [4], and then the strength generally decreases with further displacement. The following mechanisms can lead to strength reductions with displacement within GCLs and at their interfaces:

- bentonite particle orientation parallel to the direction of shear (i.e., the formation of slickensides in the bentonite);

- reinforcement failure in reinforced GCLs (i.e., fibers breaking or pulling out from carrier geotextiles);
- geomembrane polishing for both smooth and textured geomembranes;
- fiber breakage and pull-out for non-woven geotextiles; and
- weave alignment in the direction of shear for woven geotextiles.

Large-displacement strengths can be less than one-half peak strengths in some cases [4].

It is important to obtain the entire relationship between strength and displacement in laboratory testing because both peak and large-displacement strengths should be considered in design. Displacements as large as 100 mm or more may be required to reach the point at which strength no longer decreases with further displacement (i.e., the residual strength). However, laboratory test data for GCLs indicate that most of the post-peak strength loss occurs within the first 40 mm of displacement. Therefore, it is recommended that the maximum displacement be greater than or equal to 40 mm in laboratory shear testing. Preferably, the shearing stage should not be terminated until the residual strength is reached.

NORMAL STRESS

It is recommended that the shear strength be evaluated at a minimum of three different normal stresses for the following reasons. First, the relationship between strength and normal stress is not necessarily linear. For example, Fig. 7 shows that the ratio of strength to normal stress for an unreinforced GCL decreases with increasing normal stress. A minimum of three points is required to define a non-linear failure envelope. Second, the failure surface location may change with changing normal stress. In liners, the normal stress may vary due to varying waste thicknesses. In covers, the effective normal stress may vary due to variations in pre water pressures. Finally, there is variability in test results even at the same normal stress (e.g., Fig. 7). Coefficients of variation (i.e., the standard deviation divided by the mean value) from repeated direct tests at the same normal stress are presented in Table 2 for the internal strength of an unreinforced GCL and the interface strength between a non-woven geotextile and a textured geomembrane. Note that the coefficients of variation tend to be larger at lower normal stresses, and can be as large as 14 percent. By performing three tests (albeit at different normal stresses), there will be less variability in the average strength envelope than in a strength estimated from an individual test result.

Table 2--Variability in measured peak strengths from repeated tests.

Material (Normal Stress)	Number of Tests	Coefficient of Variation in Measured Peak Strength
Unreinforced GCL (18 kPa)	3	10%
Non-Woven Geotextile/Textured Geomembrane (16 kPa)	5	14%
Non-Woven Geotextile/Textured Geomembrane (340 kPa)	10	4%
Non-Woven Geotextile/Textured Geomembrane (690 kPa)	5	2%

SPECIMEN SIZE

Specimen size is generally not a significant concern for GCL strength testing, and both the smaller specimen sizes associated with ASTM D3080 sizes (on the order of 50 mm by 50 mm) and the larger specimen sizes associated with ASTM D5321 (on the order of 305 mm by 305 mm) can be used. Common practice has been to rely solely on the larger specimen sizes; however there are advantages and disadvantages associated with both smaller and larger sizes.

The main advantage with smaller specimen sizes is that direct shear tests are easier to perform. Specimen preparation, hydration and shearing all require less effort with smaller specimen sizes. In addition, conventional direct shear devices for soils are more readily available and less expensive than large-scale devices; therefore, it is easier to conduct multiple tests and perform sensitivity studies with the smaller specimen sizes. The main disadvantage with smaller specimen sizes is that the maximum possible shear displacement is limited to 5 to 10 mm. However, a torsional ring shear device with a cylindrical specimen provides a means for developing unlimited, continuous displacements with small specimens. Larger displacements in a conventional direct shear box can also be roughly simulated by shearing in 5 to 10-mm increments of displacement and resetting the upper box after each increment. Another disadvantage with smaller specimens is that the failure surface location will be forced between the upper and lower shear boxes. The shear device can be modified, such as replacing the boxes with plates, to minimize this problem. Finally, GCLs reinforced with stitches spaced at 100 mm cannot be adequately represented in smaller specimens.

The main advantage of the larger specimen size is that continuous displacements up to between 50 and 75 mm can be accommodated. Also, the failure plane is free to form anywhere in the specimen because confinement effects at the specimen boundaries are minimized due to the large specimen size. The main difficulties in performing a large-scale direct shear test with GCLs are associated with GCL hydration and drainage during shear. Care is required to provide freely draining boundaries along the GCL because horizontal drainage is limited by the large specimen size.

In conclusion, small (50 mm by 50 mm) and large (305 mm by 305 mm) specimen sizes for direct shear testing with GCLs should be viewed as complementary rather than competing alternatives. It is recommended that both specimen sizes can and should be used.

SUMMARY

The following test guidelines are recommended for measuring the internal or interface strength of a GCL in a direct shear test:

1. The GCL should be hydrated before its sheared to simulate typical field conditions.
2. When testing GCL interfaces, the interface materials (geosynthetics or soils) should be placed in contact with the GCL during hydration because bentonite extrusion from the GCL may affect the interface strength.
3. For unreinforced GCLs, the normal stress that will be applied during shear should be applied to the GCL specimen before it is hydrated. For reinforced GCLs, the order of hydration and normal stress application should model the expected field conditions.
4. Tap water should generally be used to hydrate the GCL because it is readily available and likely to be representative of typical hydration fluids in the field.
5. Care is required to provide "free" drainage to the bentonite in the GCL so that it can become hydrated quickly and completely.
6. The thickness of the GCL should be monitored during hydration. The hydration stage should not be terminated until the GCL thickness reaches a constant value with time. Incomplete hydration may produce unconservative strengths and test results that are difficult to interpret because the GCL will continue to hydrate during shear.
7. The rate of shear should be slow enough to allow for drainage of the bentonite, in accordance with ASTM D3080. Test results at faster rates are difficult to interpret due to partially drained pore water pressures in the specimen. Faster rates also appear to produce unconservative strengths.
8. The failure surface should be free to form at any location through the specimen because the measured strength depends on the failure surface location. The failure surface location should be documented.
9. The maximum displacement in the shear test should be at least 40 mm because both internal and interface strengths tend to decrease with increasing displacement after the peak strength is mobilized. Preferably, the shearing stage should not be terminated until the residual strength is reached.
10. The shear strength should be determined at a minimum of three different normal stresses to account for variability in test results and non-linearity in the failure envelope.
11. The specimen size is generally not a critical factor, and both conventional (approximately 50 mm by 50 mm) and large (305 mm by 305 mm) specimen sizes can be used. Specimen sizes for stitch-bonded GCLs should be large enough to include several rows of stitches.

It is hoped that implementation of these recommendations will reduce much of the existing scatter in measured shear strengths for GCLs and provide test results that can be interpreted without ambiguity. Also, it is hoped that these recommendations will produce strengths that are representative of those available in typical field conditions. Future research in GCL shear strength testing should focus on rate effects for both bentonite and geosynthetic components, and the interaction between bentonite and geosynthetics.

ACKNOWLEDGMENTS

The information presented in this paper has been funded in part by the United States Environmental Protection Agency under cooperative agreement CR-821448-01-0. The EPA's Project Officers on this project were Robert E. Landreth, John Martin, and David A. Carson. This paper has not been subjected to the Agency's peer and administrative review. The findings do not necessarily reflect the views of the Agency. The authors also thank the GCL manufacturers for supplying materials for testing.

REFERENCES

[1] Daniel, D. E. and Shan, H.-Y., "Results of Direct Shear Tests on Hydrated Bentonitic Blankets," *Project Report*, Geotechnical Engineering Center, The University of Texas at Austin, Texas, 1991.

[2] GRI, "Progress Report #3 - Prefabricated Bentonite Clay (PBC) Liner Test Program," Geosynthetics Research Institute, Drexel University, Philadelphia, Pa., 1991.

[3] Geosyntec, "Final Report, Direct Shear Testing, Internal Strength of Bentomat GCL," Geosyntec Consultants, Boca Raton, Florida, 1993.

[4] Gilbert, R. B., Fernandez, F. F. and Horsfield, D. W., "Shear Strength of a Reinforced Geosynthetic Clay Liner," *Journal of Geotechnical Engineering*, ASCE, in press, 1996.

[5] Daniel, D. E., Shan, H.-Y., and Anderson, J. D., "Effects of Partial Wetting on the Performance of the Bentonite Component of a Geosynthetic Clay Liner," *Proc. Geosynthetics '93*, Vancouver, B. C., 1483-1496, 1993.

[6] Schiffman, R. L., "Consolidation of Soil Under Time Dependent Loading and Varying Permeability," *Highway Research Board Proc.,* No. 37, 584-617, 1957.

[7] Shan, H.-Y. , "Stability of Final Covers Placed on Slopes Containing Geosynthetic Clay Liners," Ph.D. Dissertation, University of Texas, Austin, Texas, 1993.

[8] Mitchell, J. K., Fundamentals of Soil Behavior, Second Edition, John Wiley and Sons, New York, 1993.

[9] Gibson, R. E. and Henkel, D. J., "Influence of Duration of Tests on 'Drained' Strength," Geotechnique, Vol. 4, No. 1, 6-15, 1954.

Hydraulic Conductivity Testing Compatibility Issues

James A. McKelvey, III[1]

GEOSYNTHETIC CLAY LINERS IN ALKALINE ENVIRONMENTS

REFERENCE: McKelvey III, J. A., **"Geosynthetic Clay Liners in Alkaline Environments,"** *Testing and Acceptance Criteria for Geosynthetic Clay Liners, ASTM STP 1308,* Larry W. Well, Ed., American Society for Testing and Materials, 1997.

ABSTRACT: The use of geosynthetic clay liners (GCLs) as secondary barrier layers in environmental applications such as landfills and other impoundment facilities is becoming increasingly more popular among the engineering community, particularly at project sites where earthen materials suitable for barrier layers may not be locally available. Design engineers for these environmental applications are becoming well versed at performing equivalency calculations comparing the performance of geosynthetic materials to their earthen counterparts. For barrier layers, these equivalency calculations would normally compare the mechanical and hydraulic properties of the GCL to a compacted clay liner. Of these properties, the ability of the hydraulic properties to withstand degradation due to permeation of contained leachates is of prominent concern. Such is the case in alkaline environments. The leachate may adversely affect the GCL by minimizing swelling, decreasing adsorption capacity and increasing the permeability of the material. If the effect on the material is significant, the usefulness of this product is diminished, possibly voiding any equivalency comparison to compacted clay liner performance. The design engineer must fully understand what effect, if any, specific leachates will have on the GCL being considered. Accordingly, appropriate performance testing with the leachate in question must be performed during the design phase and confirmed during construction through quality assurance testing.

This paper will present the design considerations, required laboratory testing and conformance tests for a recent project that contained an alkaline leachate. Through appropriate testing, a contaminant resistant GCL was shown to possess desired hydraulic properties in the presence of the alkaline leachate.

KEYWORDS: cation exchange, compatibility, geosynthetic clay liners, montmorillonite, quality assurance testing

---

[1]Project Engineer, Roy F. Weston, Inc., West Chester, PA 19380.

(Now with GeoSyntec Consultants, Huntington Beach, CA 92648)

At a former copper mine in Northern California, copper and other precious metals existed along with a significant sulfur deposit in the bedrock. Through groundwater percolation and leakage into the mine workings, air and water came in contact with the sulfide minerals and other metals, enabling a chemical reaction to occur, generating a liquid commonly called acid mine drainage (AMD). The AMD at this site contains high concentrations of various metals and has a very low pH, approaching zero standard units. Unmanaged flow of AMD into the adjacent creeks have left these creeks essentially lifeless in some locations due to the extremely low pH values and high concentration of metals. Downstream from the site, the creeks converge to form a tributary to the Sacramento River. Discharge of this tributary could pose a serious threat to the aquatic environment of the river, particularly in drought conditions where dilution of the AMD is minimized. The solution selected for remediation of this site was to collect the AMD from within the mine workings and convey it to a new treatment plant where lime neutralization occurs. This process produces large quantities of calcium sulfate sludge and leachate that will have the constituents shown in Table 1. The constituents shown in this table represent worst-case values which were identified during pilot studies and later confirmed in sludge samples retrieved from the site. The sludge leaving the treatment plant has a low solids content and needs to be dewatered in drying beds. Due to high concentrations of metals in the sludge, the drying beds were required to have a composite lining system consisting of a geomembrane underlain by a low permeability soil layer.

Parameter	Average Value	Unit
pH	9.1	-
TSS	15	mg/L
TDS	3,015	mg/L
TS	3,010	mg/L
Sulfate, total	1,850	mg/L
Aluminum, total	2.00	mg/L
Calcium, total	525	mg/L
Cadmium, total	0.008	mg/L
Copper, total	0.215	mg/L
Iron, total	10.5	mg/L
Magnesium, total	28.5	mg/L
Zinc, total	0.790	mg/L

Table 1- Leachate characteristics.

Cost analyses performed during the design phase suggested that the costs associated with delivery and placement of a 9.5-cm thick compacted clay liner having a permeability less than 1 x 10^{-7} cm/s would exceed the delivery and installation costs of a GCL by approximately \$5/m^2. This indicated that a GCL would be economically superior to a compacted clay liner as the lower component of the composite liner system beneath the sludge drying beds at this site. However, the leachate characteristics shown in Table 1 suggest that there could be a compatibility problem between the leachate and the sodium montmorillonite GCLs desired for use. A clear understanding of the clay mineralogy and how this leachate reacts with the clay was essential to ensure that the design intent was achieved.

CLAY MINERALOGY

GCLs manufactured in the United States are normally produced from sodium bentonite. Bentonite is a clay material which has been formed from volcanic ash, composed largely of montmorillonite (Holeman, 1965), a mineral belonging to the smectite group. Smectites, consist of either a magnesium or aluminum octahedral sheet sandwiched between two layers of silica tetrahedra sheets which form the clay mineral's unit cell. The octahedral sheets in montmorillonites are mainly aluminum, which are often referred to as gibbsite sheets within the clay mineral structures (Mitchell, 1976). With ideal silica and gibbsite sheets, electrical neutrality is obtained by the bonding of the three layers. However, in actual clay minerals, the cations of the individual layers have been partially substituted during formation of the mineral by other cations, although the crystalline structure has remained intact.

This concept, called isomorphous substitution, leaves the clay particles with a net negative charge. Cations are therefore "...attracted and held on the surfaces and the edges, and in some clays [like montmorillonite], between the unit cells. These cations are termed 'exchangeable cations' because in most instances cations of one type may be replaced by cations of another type" (Mitchell, 1976). In montmorillonites mined in the United States for GCLs, aluminum cations have typically been partially substituted for silicon in the silica layers of the unit cells and magnesium cations have been partially substituted in the gibbsite sheets (Doheny, 1993). These substitutions are largely unbalanced, providing the high cation exchange capacity associated with montmorillonites. The sodium component of sodium montmorillonite is the exchangeable cation attached to the unit cells. GCLs manufactured in Europe typically have calcium as the exchangeable cation. The exchangeable cation in montmorillonites significantly influences the behavior of the clay minerals. Differences in the behavior of sodium and calcium montmorillonites are largely based on how the exchangeable cations react with or to other cations, water, hydroxls and other anions.

Prior to adsorption of water into the interlayer (i.e., between clay particles) of the clay particles, the thickness of the particles are approximately 12.5 and 15.4 angstroms for sodium and calcium montmorillonites, respectively (Doheny, 1993). On the basis of particle size alone, the calcium montmorillonite would be expected to have a higher permeability. However, montmorillonites have an outstanding capability of adsorbing water and other organic (dipolar) liquids through hydrogen bonding, ion hydration. osmotic attraction or dipole attraction (Mitchell, 1976). This adsorption occurs within the interlayer and along the edges of the mineral, allowing the clay particle lattice to swell. Due to the greater bonding action of calcium cations, the amount of swelling is limited when compared with montmorillonites having sodium as the exchangeable cation. Furthermore, calcium cations enhance the development of oriented water within the interlayer, leading to flocculation of the particles. Sodium cations on the other hand, "...exert little attractive force between the particles...[permitting the adsorbed water layers] to develop to very great thicknesses " (Grim, 1962). Accordingly, average free-swell for sodium montmorillonites are on the order of 1,400 percent while calcium montmorillonites swell on the average of 85 percent (Grim, 1962). The amount of swelling of the clay particles accounts for the higher liquid limit and lower permeability of sodium montmorillonite over that of calcium montmorillonite.

One drawback of GCLs produced from sodium montmorillonite used in this application is that the sodium cation is replaceable due to the small attractive force between the cation and the clay particles. This explains the very high cation exchange capacity of this clay. When exposed to an abundance of calcium, the exchange of calcium for sodium will occur readily, which would drastically change the properties of the clay (Grim, 1962). Calcium cations on the other hand require greater energy to dislodge, and therefore the cation exchange capacity is considerably lower than sodium montmorillonite. By inspection of the constituents of the leachate shown in Table 1, it can be expected that sodium montmorillonite GCLs will undergo cation exchange, and that the GCL will ultimately behave more as a calcium montmorillonite, provided there are no other compatibility concerns.

COMPATIBILITY CONCERNS

The leachate parameters shown in Table 1 posing the greatest concern for GCL compatibility are the high levels of calcium, sulfate and total dissolved solids (TDS). The effect of calcium abundance on the sodium montmorillonite was previously discussed. The magnesium cations will also exchange with the sodium cations of the clay, but may themselves be replaced by the more abundant calcium cations (Holeman, 1965). While cation exchange will occur with the other cations shown in Table 1, the small quantities of these cations will have a negligible effect on the clay. Anion exchange also occurs in montmorillonites, but to a lesser extent due to the net negative charge of the clay particles resulting from isomorphous substitution. Anion exchange rates typically average 30 milliequivalents per one hundred grams (meq/100 g) for sodium montmorillonite mined in the Northwestern United States (Doheny, 1993) compared to cation exchange rates of 80 to 150 meq/100 g (Mitchell, 1976). Accordingly, exchange of the sulfates of the leachate shown in Table 1 for the hydroxls at the edges of the gibbsite sheets will probably occur. This exchange will further reduce the negativity of the clay particles, increasing the cation exchange capacity. The high TDS will allow cation and anion exchange to begin upon hydration, as most of these ions are already in solution.

The alkalinity of the leachate may also alter the behavior of the clay, primarily by providing further cation exchange. In high pH solutions, the hydroxyls exposed on the surfaces and edges of the clay particles will have a tendency to dissociate, allowing hydrogen atoms to go into solution (Mitchell, 1976). This will leave the clay particles with a further negative net charge, increasing the cation exchange capacity. However, it was determined in this case that the influence on the clay by the pH of 9.1 would be negligible. Upon review of the aforementioned compatibility issues, it became obvious that the leachate shown in Table 1 will affect the sodium montmorillonite. The remaining question during design was whether the GCL will be able to be able to perform its function as the secondary barrier layer component of the composite lining system for the design life of the facility.

DESIGN CONSIDERATIONS AND LABORATORY TESTING

The lining system underlying the sludge drying beds at this facility was designed as a composite lining system consisting of a scrim-reinforced geomembrane underlain by a sodium bentonite GCL. In order to obtain regulatory acceptance of hydraulic equivalency to a 9.5-cm thick compacted clay liner with a permeability less than 1×10^{-7} cm/s, the GCL would have to have an initial permeability of less that 1×10^{-9} cm/s. Furthermore, regulatory acceptance required a demonstration that the hydraulic properties of the GCL would not degrade readily by advective flow of leachate through the GCL throughout the thirty year design life of the facility. While ion diffusion through the GCL would also contribute to degradation of the GCL, consideration of this transport mechanism was not required by the regulatory agencies. As the GCL was the lower component of a composite lining system, hydration of the GCL by the impounded leachate would only occur at seam defects and primary geomembrane puncture locations. Although large punctures through the geomembrane were considered unlikely due to the scrim reinforcement of the geomembrane, a 20-millimeter hole was assumed to have been placed through the geomembrane during construction of the overlying leachate collection system. The maximum hydraulic gradient occurring in the drying beds during operations was calculated to be 8.64 (McKelvey, 1995). This gradient was used in Equation 1 to determine the period required for complete cation exchange to occur in the GCL advective flow.

$$t = \frac{\rho \, CEC}{10 \, ki\Sigma I_R} \qquad (1)$$

In this relationship, which considers unit reduction, t is time (s), ρ is the mass per area of the GCL (g/cm^2), CEC is the cation exchange capacity of the clay (meq/100 g), I_R is the charge per liter of leachate (meq/L), k is the permeability of the GCL (cm/s) and i is the hydraulic gradient (dimensionless). The flow rate through the geomembrane was calculated using Darcy's Law rather than the liner leakage equations proposed by Bonaparte, et. al. (1989) as the maximum hydraulic head anticipated above the lining system would greatly exceed the thickness of the GCL, invalidating these relationships. The charge per liter of leachate passing through the GCL is determined using Equation 2.

$$I_R = X(\frac{\text{valence}}{\text{AMU}}) \tag{2}$$

where: X is the ion concentration (mg/L), valence is the electronic charge of the ion (equivalents) and AMU is the atomic mass of the ions (g). Dimensional analyses of Equations 1 and 2 are presented in the Appendix. Anion exchange is computed in a similar fashion, substituting the anion exchange capacity (AEC) for the CEC in Equation 1. The anion charge per liter due to an elevated pH is calculated by first converting the pH value to a moles/liter unit and then determining the anion exchange due to the hydroxls in solution within the leachate. If it is assumed that cation exchange and anion exchange occur concurrently, the time required for equilibrium to occur can be assumed to be the greater of the periods required for cation and anion exchange.

The time required for the leachate shown in Table 1 to cause total cation and anion exchange to occur within the GCL considering only advective flow was calculated using Equations 1 and 2 assuming an anion exchange capacity of 30 meq/100 g, a cation exchange of 115 meq/100 g and a constant hydraulic gradient. The permeability values used in the analysis ranged from 2.5×10^{-7} cm/s to 1.5×10^{-9} cm/s, typical of calcium and sodium montmorillonites, respectively (Grim, 1962). The greater permeability would permit equilibrium in just a few years, while the lower permeability would require a period for total cation and anion exchange to occur far in excess of the design life of the facility. These calculations were considered conservative as initial hydration of the GCL would occur through suction of subgrade moisture into the GCL (Daniel, et.al, 1993) rather than by leachate hydration and that the hydraulic gradient used would not occur over the entire exchange period. However, as the permeability of the sodium montmorillonite would increase during the exchange period, compatibility testing of this leachate with various GCLs was initiated.

The swelling capability of the bentonite used in GCLs is easily evaluated by performing free swell tests in accordance with USP-NF XVII, which measures the amount of swelling of 2 grams of clay when placed in a graduated cylinder containing 100 mL of distilled water. This test could also evaluate the initial effects the leachate has on the swelling of the bentonite by substituting leachate for the distilled water, and comparing the results to the standard test. Figure 1 shows that the initial swelling capability of two candidate sodium bentonite GCLs was severely diminished when tested in the leachate, indicating that the ion exchange would occur readily during hydration. The higher permeability assumed in the design calculations would likely occur, which was undesirable. Several GCL manufacturers were consulted at this time. It was recommended that a contaminant resistant sodium bentonite be evaluated for this leachate. The underlying premise behind contaminant resistant sodium bentonite is that a sacrificial additive is included within the bentonite which promotes ion exchange with the additive rather than with the exchangeable ions of the clay. How this interaction occurs is not understood due to the proprietary nature of the additives. As also shown in Figure 1, contaminant resistant sodium bentonite performed quite well with the leachate in the free swell tests.

Further compatibility testing was performed using the filtrate volume or fluid loss test described by API 13A. This test, which is less susceptible to human error involves preparing a suspension of bentonite and liquid which is then placed in a filter press cell and monitored for filtrate volume. This test provides a loose indication of permeability. Figure 2 presents results of fluid loss testing performed on standard grade and contaminant resistant sodium bentonite using distilled water, an aggressive 1,000 ppm calcium chloride solution and the leachate shown in Table 1. Based on the results of free swell and fluid loss tests, it was concluded that a contaminant resistant bentonite would be required to provide the necessary swelling during hydration in order to provide the initial low permeability desired to extend the ion exchange period.

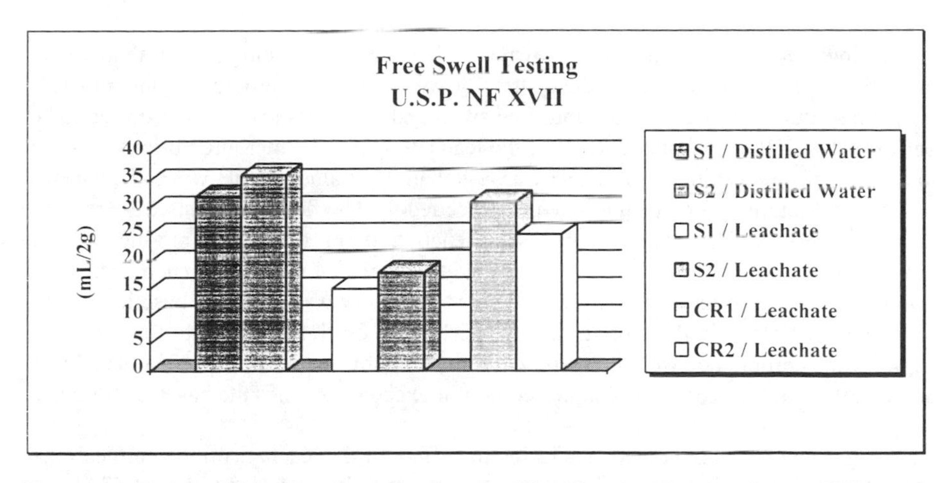

Figure 1 - Free swell testing of standard grade (S) and contaminant resistant (CR) sodium montmorillonites. The leachate shown in Table 1 was used for the leachate free swell tests.

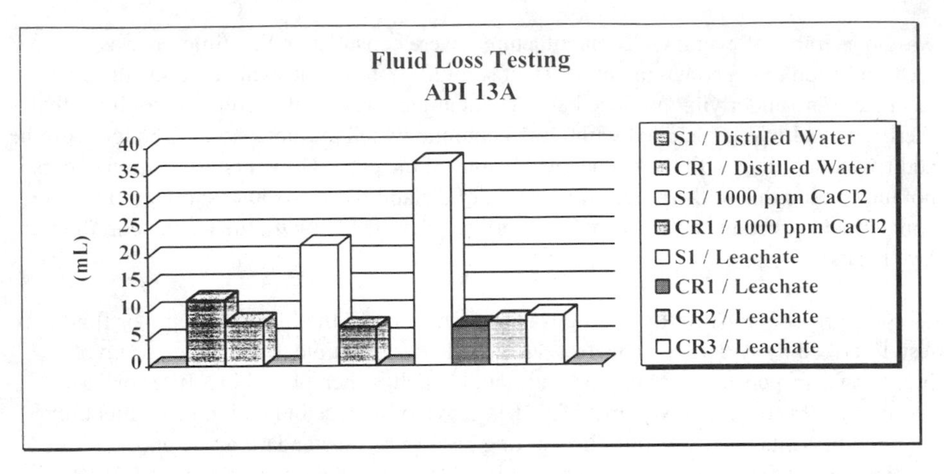

Figure 2 - Fluid loss testing of standard grade (S) and contaminant resistant (CR) sodium montmorillonite hydrated with distilled water, a 1000 ppm calcium chloride solution and the leachate shown in Table 1. Three test results are shown for the contaminant resistant clay hydrated with leachate.

Quality Assurance

In order to provide the facility owner with a composite lining system that would meet all of the design objectives, the contractor was required to provide a contaminate resistant sodium bentonite GCL. Both free swell and fluid loss tests were required to be performed with the leachate. The pH of the leachate was to be measured prior to testing to ensure a pH exceeding 9. If the pH was less than this value, the pH was to be adjusted by adding calcium hydroxide until the pH exceeded 9. The free swell tests were required to be performed at a frequency of one test per each 2,800 m^2 of material produced for this project. The amount of free swell was required to exceed 24 mL/2 g for each test. The average of four fluid loss tests performed prior to delivery was used as a baseline for field conformance tests. Six fluid loss tests were performed on field samples to further demonstrate compliance with the specification. Accounting for variability in the additive process, the fluid loss of these samples could not exceed 1.2 times the baseline fluid loss.

In order to quantitatively determine the effect of the leachate on the permeability of the GCL, five flexible-wall permeability tests were to be performed using ASTM D5084. For regulatory acceptance, the GCL was required to be evaluated without initial hydration and performed under an effective confining stress of 34.5 kPa using the leachate as the permeant. Ideally, this testing would have lasted until a least one pore volume of leachate passed through the samples; a period requiring approximately two months, but due to time constraints of the project, were limited to 14 days. Figure 3 shows the results of one of the tests, which indicates that the permeability after hydration is lower than the 1 x 10^{-9} cm/s value required.

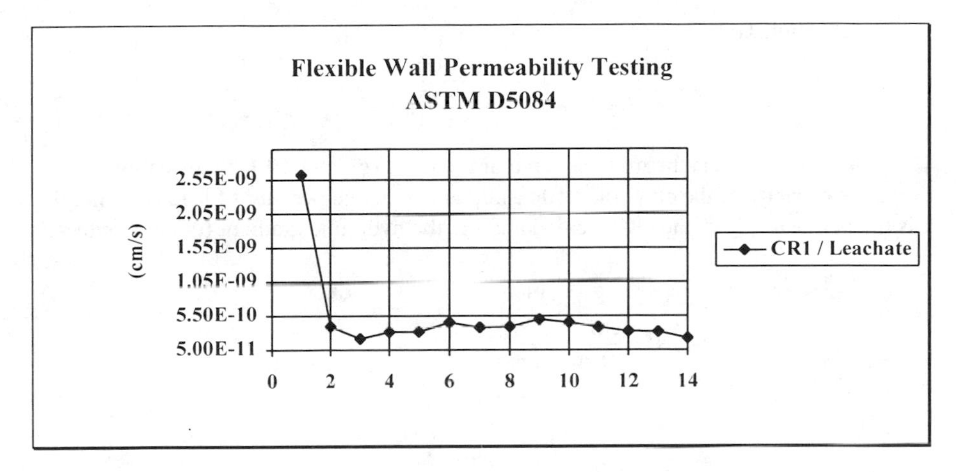

Figure 3 - Fourteen day permeability test result of contaminant resistant GCL with leachate as permeant. Test was performed without initial hydration and under an effective confining stress of 34.5 kPa.

SUMMARY

Sodium bentonite GCLs will undergo extensive ion exchange when permeated with alkaline leachates containing high percentages of calcium and other ions. If standard grade sodium bentonites are used in this application, swelling of the clay particles will be reduced during hydration due to ion exchange, leading to a higher permeability and a shorter ion exchange period. A qualitative assessment of the ion exchange occurring during hydration of the GCL can be made by performing simple index tests such as free swell and fluid loss tests. The results of this testing may indicate that sodium bentonites containing sacrificial additives may be required to provide adequate swelling during hydration. Through index testing and permeability testing, it is possible that adequate long- term performance could be demonstrated mathematically through ion exchange calculations. A quantitative assessment of permeability versus ion exchange may be warranted for particularly aggressive leachates.

Further research is needed to develop ion exchange permeability relationships. The paper also suggests the need for standardized procedures for free swell and fluid loss index testing. Laboratory test procedures for determining ion exchange capacity of bentonites would also be helpful.

APPENDIX

As the presented equations include unit reductions, they may at first glance appear dimensionally incorrect. Accordingly, the following dimensional analyses are presented.

Equation 1:

$$t = \frac{\rho\, CEC}{10\, ki\Sigma I_R}$$

t is time (s), ρ is the mass per area of the GCL (g/cm^2), CEC is the cation exchange capacity of the clay (meq/100 g), I_R is the charge per liter of leachate (meq/L), k is the permeability of the GCL (cm/s) and i is the hydraulic gradient (dimensionless).

$$s = \frac{\frac{g}{cm^2}\ \frac{meq}{100g}}{\frac{10cm}{s}\ \frac{meq}{L}}$$

$$= \frac{\frac{meq}{100cm^2}}{\frac{10cm}{s}\ \frac{meq}{L}\left(\frac{L}{1000cm^3}\right)}$$

$$= \frac{\frac{meq}{100cm^2}}{\frac{meq}{s\ 100cm^2}}$$

$$s = s \qquad \text{OK.}$$

Equation 2:

$$I_R = X\left(\frac{valence}{AMU}\right)$$

X is the ion concentration (mg/L), valence is the electronic charge of the ion (equivalents) and AMU is the atomic mass of the ions (g).

$$\frac{meq}{L} = \frac{mg}{L}\left(\frac{eq}{g}\right)$$

$$= \frac{mg}{L}\left(\frac{eq}{g}\right) x \left(\frac{1000meq}{eq}\right) x \left(\frac{g}{1000mg}\right)$$

$$\frac{meq}{L} = \frac{meq}{L} \qquad \text{OK.}$$

REFERENCES

Bonaparte, R., Giroud, J.P. and B.A. Gross, (1989) "The Rates of Leakage Through Landfill Liners", *Proc. of Geosynthetics '89*, San Diego, CA, IFAI, St. Paul, pp. 18-29.

Daniel, D.E., Shan, H-.Y. and J.D. Anderson, (1993) "Effects of Partial Wetting on the Performance of the Bentonite Component of a Geosynthetic Clay Liner", *Proc. of Geosynthetics '93*, Vancouver, B.C., Canada, IFAI, St. Paul, pp. 1483 - 1496.

Doheny, E.J., (1993) "Advanced Engineering Geology", Course notes, Drexel University, Philadelphia.

Grim, R.E., (1962) Applied Clay Mineralogy, McGraw-Hill, New York.

Holeman, J.N., (1965) "Clay Minerals", Technical Release No. 28, U.S. Department of Agriculture.

Mitchell, J.K., (1976) Fundamentals of Soil Behavior, Wiley, New York, 422 p.

McKelvey, J.A., (1995) "Filtration System Design for Sludge Drying Beds", *Proc. of Geosynthetics '95*, Nashville, TN, IFAI, St. Paul.

Mlynarek, J.[1], Vermeersch, O.G.[1], and Lemelin, D.[2]

LABORATORY SIMULATION OF GEOSYNTHETIC CLAY LINER APPLICATION IN CONTAMINATED LIQUIDS EVACUATION

REFERENCE: Mlynarek, J., Vermeersch, O. G., and Lemelin, D., **''Laboratory Simulation of Geosynthetic Clay Liner Application in Contaminated Liquids Evacuation,''** *Testing and Acceptance Criteria for Geosynthetics Clay Liners, ASTM STP 1308,* Larry W. Well, Ed., American Society for Testing and Materials, 1997.

ABSTRACT : To prevent a contamination of soil and underground water by leaking mineral oil, recovery basins are designed and constructed in Quebec, Canada. The functions of such basins are to collect and to evacuate oil to a drainage and then to a recycled and treatment station. The material presently used for such an application is a concrete. However, due to difficult access to some of the transformers, and to the difficult low temperature conditions, engineers are looking for a new, alternate design idea. In order to evaluate the geosynthetic clay liner (GCL) hydraulic behavior in such applications, a laboratory demonstration test has been conducted. A full-scale model was designed and constructed for the purpose of measuring the rate of water flow through different layers of the proposed system. Mineral oil leaks as well as precipitation were simulated during the research program. The testing consisted of the measurements of mineral oil and water (precipitation) volumes at four levels of the demonstration model, during a period of two months. The results showed that only one percent of precipitated water and leaked mineral oil was collected underneath the geosynthetic clay liner. Further research is recommended on :

- techniques of seaming of GCLs' joints and connections;
- the minimum acceptance rate of hydration of GCLs for different liquids;
- an influence of water content of soils on GCLs hydration;
- a long term hydraulic compatibility of GCLs with different liquids and leachates.

---

[1] General Manager and Adjunct of R&D Manager, respectively, Geosynthetics Analysis Service (SAGEOS), 3000 rue Boullé, Saint-Hyacinthe, Qc, Canada J2S 1H9.

[2] Project Engineer, Hydro-Québec, Division Ingénierie, Service Postes de répartition, 500 rue Sherbrooke O., Montréal, Qc, Canada H3A 3C6.

According to the IFAI market report (1994) about 3,900,000 m^2 of Geosynthetic Clay Liners (GCLs) were installed around the world in 1992, and the estimated surface for 1995 is 7,000,000 m^2 in different fields of applications :

- municipal waste containment systems	55%
- industrial waste containment systems	15%
- agricultural tanks	10%
- mining waste	10%
- municipal and industrial basins	10%

Among potential applications of GCLs, one presents a particular interest: the use of GCLs in mineral oil recovery basins. The function of a GCL in such an application is to avoid pollution of groundwater and soil by collecting and evacuating leaking oil to a drainage collection system.

Presently, such specific recovery basins for mineral oils are designed and constructed using a concrete material. However, in some situations this material seems to be not the most economical one. Indeed, due to the difficult access to some basins and to the difficult climate conditions, it could be less expensive and more practical to design an alternate recovery basins using geosynthetic materials.

GCLs are the materials considered for possible installation in this application. However, chemical compatibility of GCLs, and sealing properties of their seams and joints are of major importance.

Considering those facts, a research program has been developed and conducted using a full-scale laboratory demonstration chamber. The main objective of the project was to evaluate the hydraulic behavior of a commercial GCL in the application as a liner in the recovery basin.

EXPERIMENTAL PROGRAM

Simulation chamber and testing procedure

A sector of the recovery basin in 1:1 scale was built in the laboratory. A simulation chamber, shown in Fig. 1, was designed, based on the dimensions of the actual constructed basins.

The sample of GCL was installed in the chamber without preliminary hydration. Sealing of the interface of the GCL with the wall of the chamber was assured by using a 2.5 cm overlap, and a silicon adhesive. The medium sand, placed beneath the GCL, as well as the 0-20 calibrated granular drainage material, placed on the top of the GCL, were compacted to 95% standard Proctor.

Oil leaks and precipitation were simulated and controlled by using a peristaltic pump and calibrated sprinklers. Two sprinklers were attached to the top of the model. Injection of the mineral oil was simulated at the upper part of the model, as shown in Figure 1. In addition, a tracer-dye (rhodamine) was injected to visualize the distribution of water and mineral oil flowlines in the different layers of the simulation chamber.

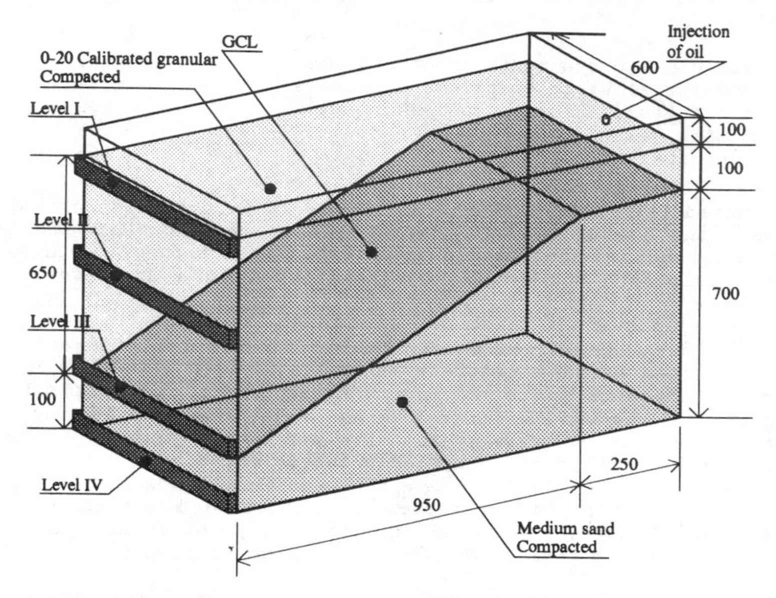

Fig. 1--Schematic of the simulation chamber

The oil leaks, as well as precipitation, were simulated twice during the program, at the beginning of the test and after ten days of testing.

Drainage collectors, located at four levels of the chamber were used to drain oil and water. The locations of the collectors are shown in Figure 1 :

- level I : at the surface of the 0-20 calibrated granular drainage material, on the top of the chamber;
- level II : in the middle of the 0-20 calibrated granular drainage material, 320 mm beneath the drainage surface;
- level III : at the GCL surface;
- level IV : underneath the GCL and the medium sand, at the bottom of the chamber.

The testing procedure consisted of the measurements of oil and water (precipitation) volumes injected into the chamber, and at the outlets of drains at each level of the model, during the two month period.

At the end of the testing program, samples of GCL as well as sand and calibrated gravel were taken from different location of the simulation chamber in order to verify their water content (level of hydration).

Properties of Tested GCL

Selected properties of the GCL, used in the research program, have been evaluated. They consist of mass per unit area (SAGEOS GC004 [1]), thickness (ASTM D5199 [2]), hydraulic conductivity (ASTM D5084 [3]), grab breaking load and elongation (ASTM D4632 [4]) and hydraulic compatibility with leaking oil (SAGEOS GC001 [5]). The results are summarized in Table 1.

It can be noted that the hydraulic compatibility of a GCL with mineral oil depends on the water content of the GCL. It appears that the non-hydrated GCL is not compatible with mineral oil (high percolation rate - 3×10^{-3} s^{-1}), whereas the partially hydrated GCL, tested under free swelling conditions (no charge applied during sample hydration), shows a quite low percolation rate - 2×10^{-7} s^{-1}.

The details of the procedure, used to determine the percolation rate (permittivity) of partially hydrated GCLs, are presented in a paper by Mlynarek *et al.* [6].

Finally, testing the completely saturated sample by using the ASTM D5084 procedure, the hydraulic conductivity of the GCL is as low as 1.7×10^{-9} cm/s (equal to a permittivity of 2.13×10^{-9} s^{-1}).

Results

The exact values of precipitation and injected leaks are reported in Table 2.

Periodic measurements of oil and water (precipitation) volumes at each drainage level of the model have been carried out during a period of two months.

The volumes collected are shown in Figures 2 to 4, and summarized in Table 4. Flow rate values are presented in Figures 5 to 7.

It should be noted that no volumes were collected at drainage levels I and II, thus Figures 2 to 4 show only volumes collected at levels III (drain located at the GCL surface drain) and IV (drain located at the bottom of the chamber).

Finally, Figure 8 shows the water content of the GCL samples at different locations in the model at the end of testing, e.g. after 62 days of testing, and after 52 days of a second precipitation simulation (this precipitation lasted one day). Water content of the soils are also shown in this Figure.

Table 1--Selected Properties of Tested GCL

Method	Property	Value	Direction	
			Machine	Cross
ASTM D4632	Max. Grab Strength	Mean (N)	2900	1084
		CV (%)	5.5	8.7
	Elon. at Max. grab strength	Mean (%)	67.4	134.7
		CV (%)	3.5	5.1
ASTM D5199	Thickness	Mean (mm)	8.0	
		CV (%)	8.2	
SAGEOS GC004	Mass per unit area	Mean (g/m^2)	4 446	
		CV (%)	13.1	
ASTM D5084	Hydraulic conductivity	Mean (cm/s)	1.7 x 10^{-9}	
		CV (%)	10.5	
	Hydraulic permittivity	Mean (s^{-1})	2.13 x 10^{-9}	
		CV (%)		
SAGEOS GC001	Percolation rate note1	(s^{-1})	3 x 10^{-3}	
	Percolation rate note2	(s^{-1})	3 x 10^{-5}	
	Percolation rate note3	(s^{-1})	2 x 10^{-7}	

Note 1: Percolation rate of non-hydrated GCL to mineral oil.
Note 2: Percolation rate of partially hydrated GCL (50% of water by weight of dry sample) to mineral oil.
Note 3: Percolation rate of partially hydrated GCL (100% of water by weight of dry sample) to mineral oil.

Table 2--Values of precipitation and oil injection

Injection	Liquid	Total	Rate
First injection	Water	234.2 liters	0.488 liters/min.
	Oil	500 grams	1.85 g/min.
	Tracer-dye	0.5 l (2800 ppm)	1.85 ml/min.
Second injection	Water	237.1 liters	0.494 liters/min.
	Oil	964 g.	2.87 g/min.
	Tracer-dye	0.573 liter (2800 ppm)	1.71 ml/min.

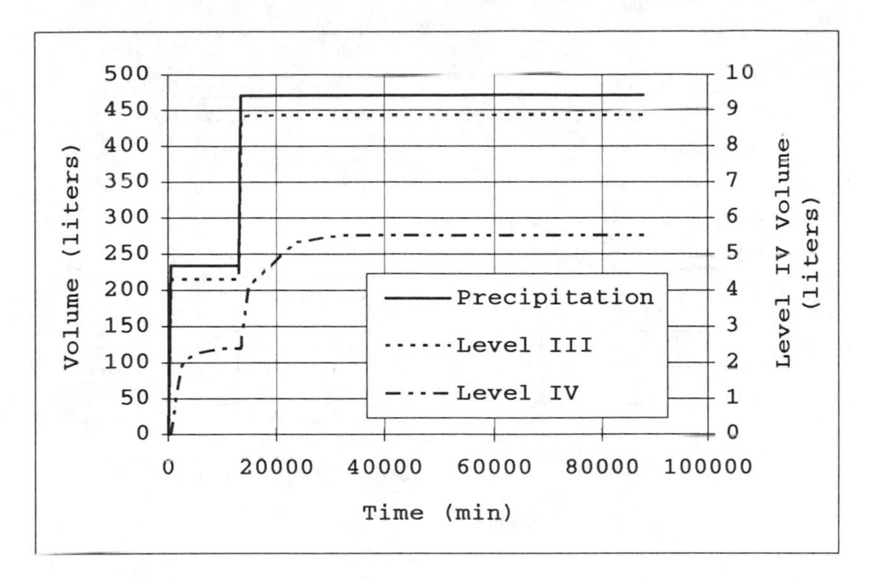

Fig.2--Cumulative curves of collected volumes of flow at levels III and IV of the chamber (complete scale)

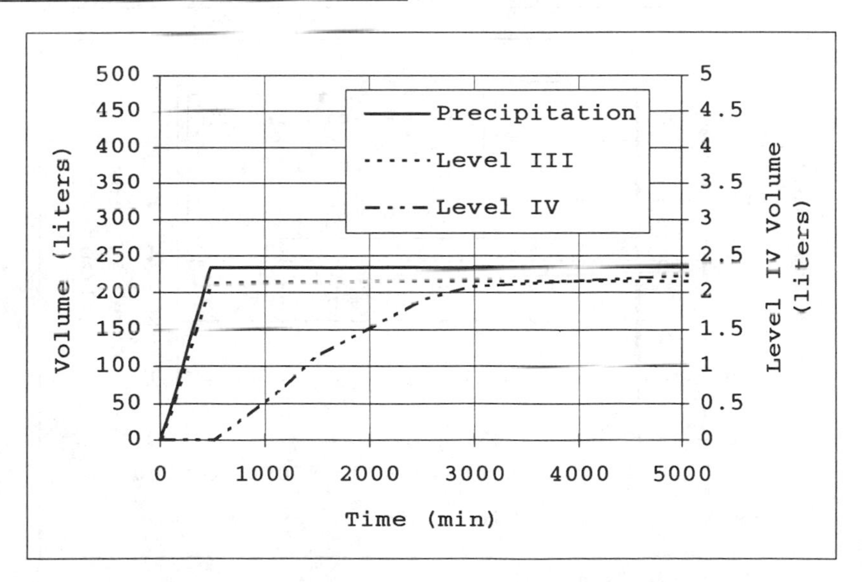

Fig.3--Cumulative curves of the collected volumes of flow at levels III and IV of tested system(detail of first precipitation simulation)

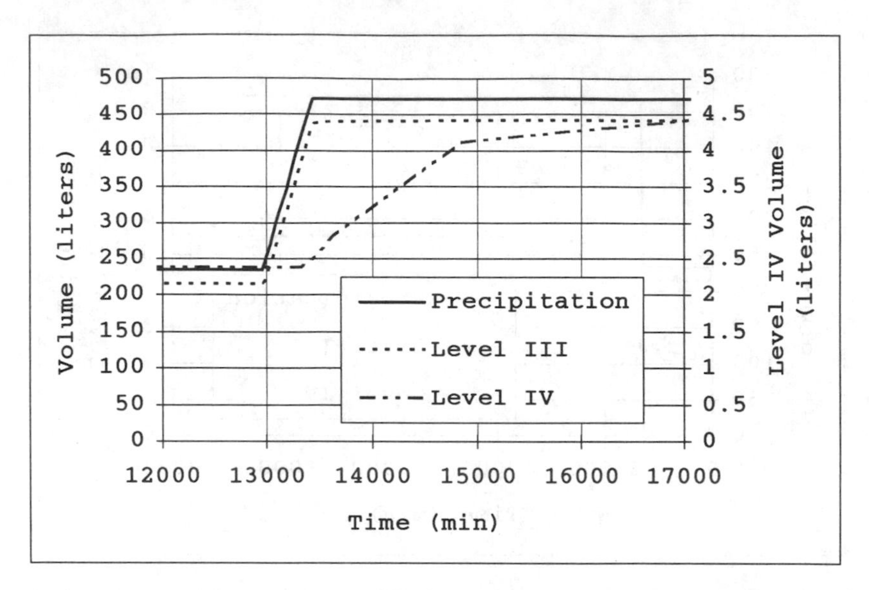

Fig.4--Cumulative curves of the collected volumes at the levels III and IV of the tested system (detail of second precipitation simulation)

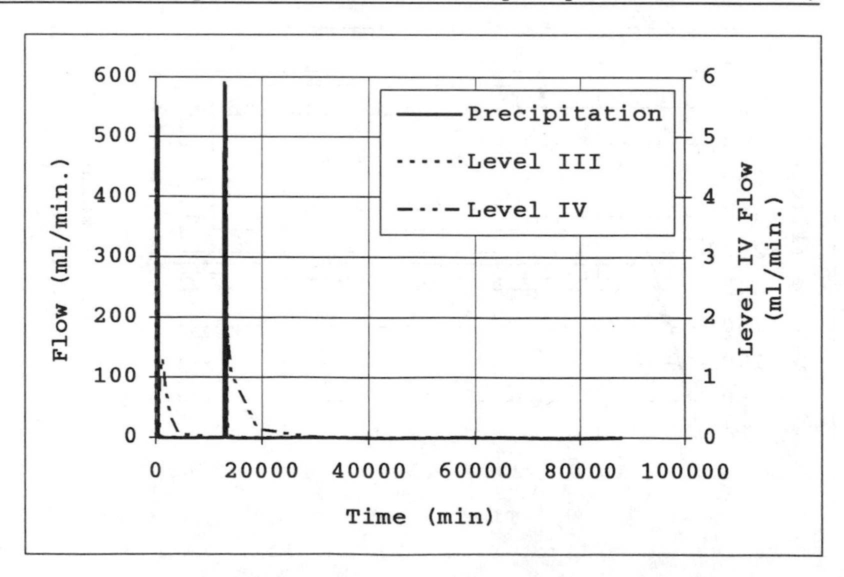

Fig.5--Flow rates from drains located at the level III and IV (complete scale)

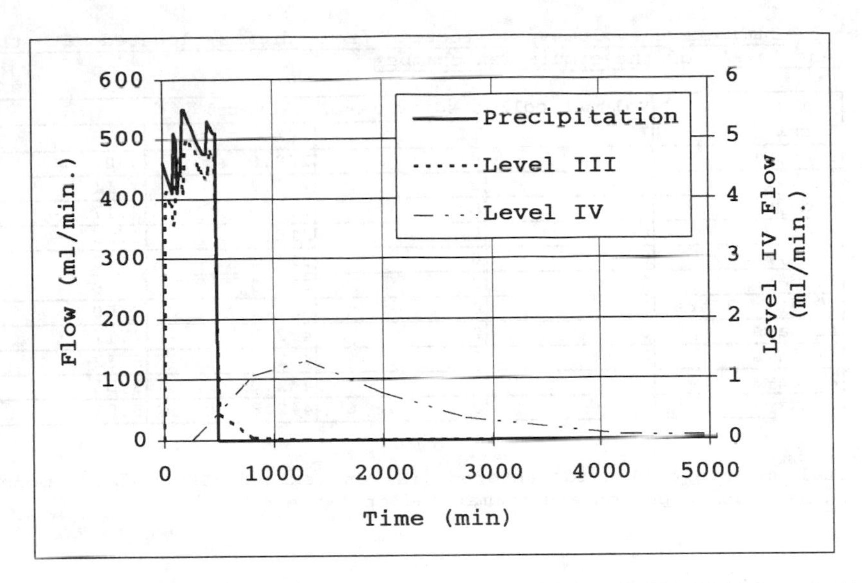

Fig.6--Flow rates from drains at the level III and IV (detail of first precipitation simulation)

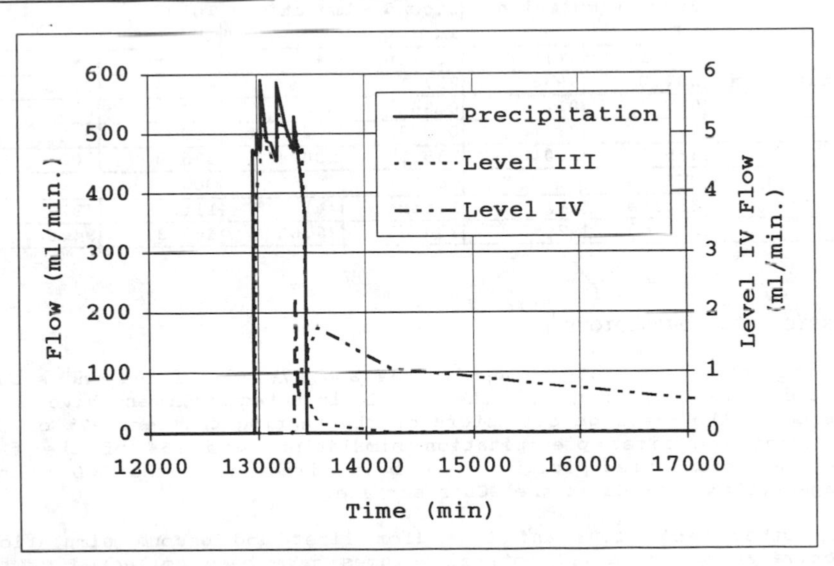

Fig.7--Flow rates from drains located at the level III and IV (detail of second precipitation simulation)

Table 3--Cumulative volumes collected from drains located at the different levels of the simulation chamber

After	Volumes collected at level (liters)			
	I	II	III	IV
8 hours	0	0	208.6	0.00
1 day	0	0	214.7	1.11
2 days	0	0	215.0	2.17
4 days	0	0	215.2	2.25
7 days	0	0	215.3	2.38
10 days	0	0	215.3	2.38
10 d. + 8 hrs	0	0	434.2	2.50
11 days	0	0	442.0	4.11
17 days	0	0	442.8	5.32
23 days	0	0	442.8	5.51
61 days	0	0	442.8	5.51

The total amounts of liquids entered (IN) and exited (OUT) of the model during the testing period are summarized in Table 4.

Table 4--Amounts of IN and OUT liquids

Liquid	First simulation IN		Second simulation IN		Total IN	
	liters	(%)	liters	(%)	liters	(%)
Precipitation	234.2	99.6	237.1	99.4	471.3	99.5
Oil	0.5	0.2	0.964	0.4	1.464	0.3
Tracer	0.5	0.2	0.573	0.2	1.073	0.2
Total	235.2	100	238.6	100	473.8	100
	OUT		**OUT**		**OUT**	
	liters	(%)	liters	(%)	liters	(%)
Total	217.68	92.55	230.63	96.66	448.31	94.6

DISCUSSION AND CONCLUSIONS

It can be noticed from Table 4 that 93% and 97% (for first and second simulations respectively) of the total injected volumes have been recovered at the drainage collectors of the testing chamber. 92% of the flow during the first precipitation simulation, and 95% of the flow during the second precipitation have been directed by the GCL to the drainage system located at the GCL's surface.

On the other hand, 1.0% and 1.3% (for first and second simulations respectively) of the total entered volumes have been collected by the drainage system located underneath the GCL, at the bottom of the simulation chamber. Nevertheless, these volumes are probably due to a leak at the interface of the GCL with the chamber wall. Visual observation of tracer-dye distribution clearly showed a passage (leaks) of the liquids along the joints between the chamber wall and the GCL. Designing and installation of the recovery basins should consider the joint between the GCL and any dissimilar material to be a weak joint. Therefore, the recommended installation technique for securing GCL to vertical wall is shown in Figure 9.

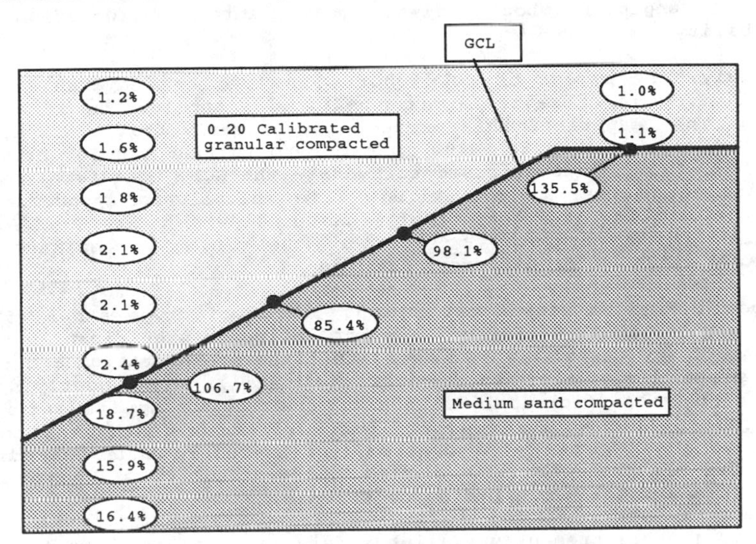

Fig. 8--Distribution of water content of the GCL and soils in the chamber at the end of testing (52 days after last precipitation simulation)

Further research and testing is also needed to develop techniques of sealing GCLs with construction elements.

The evaluation of hydraulic compatibility of a GCL with mineral oil reveals that this characteristic depends on the level of hydration (water content) of the GCL. The reported research confirms that the tested non-hydrated GCL is not compatible with mineral oil (high percolation rate - $3 \times 10^{-3}\ s^{-1}$). It was also found that in case of the non-confined GCL, a water content of 200% is needed to obtain GCL's hydraulic compatibility with the oil (percolation rate to the

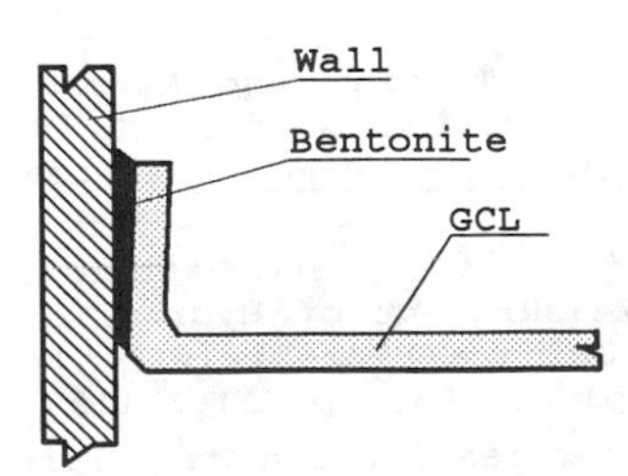

Fig. 9--Schematic design of sealing of the interface of GCLs and construction elements

oil - 2 x 10^{-7} s^{-1}). It has to be mentioned that non-confined conditions of GCL's hydration are not recommended by all GCLs` manufacturers in engineering practice. However, in some cases such conditions can occur, thus an engineer should always take into consideration such a possibility.

Obviously, the aspect of GCLs short and long term hydraulic behavior with liquids is of major importance for its application in a recovery basins. The results of this project show clearly that a GCL has to be hydrated up to a certain value in order to fulfill its low permeability function. Thus, it is proposed to hydrate the GCL, immediately after the construction works, by sprinkling. This installation procedure should avoid possibility of a contact of leaking oil with a dry GCL. For future research it is suggested to look at the minimum acceptance rate of hydration of GCLs for different liquids.

It should also be noted that a long period of dry weather (52 days), simulated during this project, did not influence significantly the water content of the tested GCL. It seems that the pre-hydration and the long-term process of water suction (Daniel [7]), from underlying soils, should assure sufficient hydration of the GCL during its service life. Research on this topic, e.g. influence of water content of underneath soils on a GCLs rate of hydration, is presently under way in SAGEOS hydraulic laboratories to better understand GCLs' water suction properties and the whole hydration process.

Finally, most of presently available data is based on short term testing (few days to few months). Thus, a long term testing of hydraulic compatibility of GCLs to different liquids and leachates are strongly recommended.

AKNOWLEDGEMENTS

The authors wish to thank Hydro-Quebec, Quebec, Canada for financing the reported project. Thanks to Mr. Mathias Kolos of Hydro-Quebec for his valuable discussions.

REFERENCES

[1] SAGEOS Geosynthetics Analysis Service, SAGEOS GC-004/95, Geosynthetics Testing Methods Mass per Unit Area of Geosynthetic Clay Liners.

[2] American Society for Testing and Materials, ASTM D5199, Standard Test Method for Measuring Nominal Thickness of Geotextiles and Geomembranes, Annual Book of ASTM Standards, Vol. 04.09, pp. 99-101.

[3] American Society for Testing and Materials, ASTM D5084-90, Standard Test Method for Measurement of Hydraulic Conductivity of Saturated Porous Materials Using a Flexible Wall Permeameter, Annual Book of ASTM Standards, Vol. 04.08, pp. 1161-1168.

[4] American Society for Testing and Materials, ASTM D4632-91, Standard Test Method for Grab Breaking Load and Elongation of Geotextiles, Annual Book of ASTM Standards, Vol. 04.08, pp. 939-942.

[5] SAGEOS Geosynthetics Analysis Service, SAGEOS GC-001/95, Geosynthetics Testing Methods - Hydraulic Compatibility of Geosynthetic Clay Liners.

[6] Mlynarek, J., Blond. E., Vermeersch, O.G., "Hydraulic Compatibility of Partially Hydrated Geosynthetic Clay Liners (GCLs) with Contaminated Liquids", Testing and Acceptance Criteria for Geosynthetics Clay Liners, ASTM STP 1308, Larry W. Well, Ed., American Society for Testing and Materials, Philadelphia, 1996.

[7] Daniel, D.E., Gilbert, R.B., "Geosynthetic Clay Liners for Waste Containment and Pollution Prevention", Proc. of Short Course, University of Texas a Arustin, 1Austin, USA, 1994.

John D. Quaranta,[1] Mohammed A. Gabr,[2] and John J. Bowders[3]

FIRST-EXPOSURE PERFORMANCE OF THE BENTONITE COMPONENT OF A GCL IN A LOW-pH, CALCIUM-ENRICHED ENVIRONMENT

REFERENCE: Quaranta, J. D., Gabr, M. A., and Bowders, J. J., ''**First-Exposure Performance of the Bentonite Component of a GCL in a Low-pH, Calcium-Enriched Environment,**'' *Testing of Geosynthetic Clay Liners, ASTM STP 1308,* Larry W. Well, Ed., American Society for Testing and Materials, 1997.

ABSTRACT: Testing was conducted on the bentonite portion of a Geosynthetic Clay Liner (GCL) for application to an environment characterized as having high concentrations of dissolved calcium ions. This environment presents conditions that might affect the long-term hydraulic function of the GCL as a component in a barrier system. Experiments were conducted to investigate first-exposure compatibility of a sodium bentonite GCL subject to the affects of acidic groundwater and second from the combined affects of acidic groundwater enriched with calcium. Relationships between the ionic exchange of sodium, potassium, magnesium and calcium species in the bentonite, and changes in hydraulic conductivity and electrical conductance are reported and discussed.

KEYWORDS: geosynthetic clay liner, hydraulic conductivity, ion exchange, bentonite, cation exchange capacity

---

Geosynthetic Clay Liners (GCLs) are geosynthetic products which use bentonite as a hydraulic barrier. The largest market for GCLs is in waste containment applications. Several types of GCLs are available; the reader is directed to Koerner (1) for a detailed discussion on GCL materials, construction, manufacturers, and design applications.

The GCL selected for testing was the GUNDSEAL® product manufactured by GSE Lining Technology, Inc. Houston, TX. GUNDSEAL® is a GCL manufactured with one pound per square foot of high quality sodium bentonite adhered to the sheet using a patented nontoxic, water soluble adhesive. This GCL may be installed such that the bentonite component is either a primary or secondary barrier.

The research presented in this paper was undertaken to investigate the first-exposure, chemical compatibility of a Geosynthetic Clay Liner (GCL) when used in an environment with high concentrations of dissolved calcium ions. This environment presents unique and challenging conditions that might affect the long-term hydraulic function of the GCL.

---

[1] Engineering Scientist, Department of Civil & Environmental Engineering, West Virginia University, Morgantown, WV 26506-6101
[2] Associate Professor, Department of Civil & Environmental Engineering, West Virginia University, Morgantown, WV 26506-6101
[3] Research Civil Engineer and Senior Lecturer, Department of Civil Engineering, University of Texas, Austin, TX 78712-1076

The experimental program was designed to approximate conditions that a GCL might experience when placed in an environment where the permeant liquid would have a low pH, so as to simulate a waste leachate or an acidic groundwater, and contain significant concentrations of dissolved cations.

The goal was to investigate the resulting hydraulic conductivity of the GCL's bentonite permeated in flexible-wall triaxial cells with two liquid permeants, acetic acid alone and acetic acid enriched with calcium. Chemical analyses of the effluents were performed to identify ion replacement occurring between the sodium, potassium, magnesium, and calcium ions in the bentonite with the enriched calcium permeant liquid, and cation replacement due to an acidic permeant liquid. Chemical assays of the sodium bentonite were performed to characterize the Cation Exchange Capacity (CEC), soluble salts ratio, pH, and electrical conductance.

MATERIALS AND METHODS

MATERIALS

BENTONITE

In the as-manufactured condition, the tested GUNDSEAL® consisted of high-quality (90% to 94% sodium montmorillonite) sodium bentonite adhesively bonded to a 1.5 mm (60 mil) smooth high density polyethylene (HDPE) liner (2). The physical properties of the grey colored bentonite were: initial water content (*w*) 17.7%, initial saturation (S) 28%, Liquid Limit (LL)=330, and Plasticity Index (PI)=282. The specific gravity (G_s) used in the determining the soil properties was 2.79 (2).

Two additional bentonite specimens were analyzed for comparison with the GUNDSEAL® bentonite. These specimens included samples of a raw, as processed bentonite (BARA-KADE® LD-16), and a treated, chemical resistant bentonite (ENVIRO-SEAL® LD-16). The samples were furnished by the BENTONITE Corporation, Denver CO.

Both specimens were granular and showed the following properties: BARA-KADE® LD-16, LL=675 and PI=625; and ENVIRO-SEAL® LD-16, LL=779 and PI=737.

PERMEANTS

In limestone deposits solubilization of calcium (Ca^{2+}) occurs naturally from contact with acidic groundwater, where concentrations of Ca^{2+} commonly range 0.1 to 0.2 M (3). Acetic acid at a pH=3.0 was selected as the permeant liquid based on information from Baver (4), and Daniel and Ruhl (5). It was also useful for comparison with results by Daniel and Ruhl (5) who tested GUNDSEAL® using permeant liquids for a simulated municipal solid waste leachate (pH=4.35) and a simulated hazardous waste leachate (pH=3.0).

The permeant liquids were prepared using distilled and deionized water as the solvent to which a 0.03M acetic acid and a 0.2M calcium carbonate solute were added making an acidic calcium acetate solution. Target and measured compositions for the permeant liquids are shown in Table 1.

Table 1 Permeant Liquid Properties

Property	Cell #1	Cell #2
Solution	Calcium Carbonate + Acetic Acid	Acetic Acid
pH at 25°C	3.0	3.0
Concentration	0.2 M Ca^{2+} + 0.03 M	0.03 M
Ca^{2+} Concentration (mg/l) at 25°C	4,920	
Electrical Conductance at 25°C	6,270 μS	286 μS

μS= microSiemens

METHODS

Apparatus

Testing was conducted using triaxial-cell permeameters in accordance with ASTM D5084. No backpressure was applied to the specimens and the bentonite specimens were not prehydrated. The initial hydraulic gradient was 50 and the cell pressure was maintained at 14 kPa (2 psi). The effluent was collected into enclosed graduated cylinders maintained at atmospheric pressure and subsequently analyzed for ion concentrations.

For all testing, intact bentonite specimens were removed from the HDPE liner using a flat-bladed knife. The adhesive allowed the specimens to maintain their integrity. Initial specimen dimensions measured 71 mm in diameter by 4 mm thick. Configuration of the specimens in the permeameter cells is shown in Figure 1.

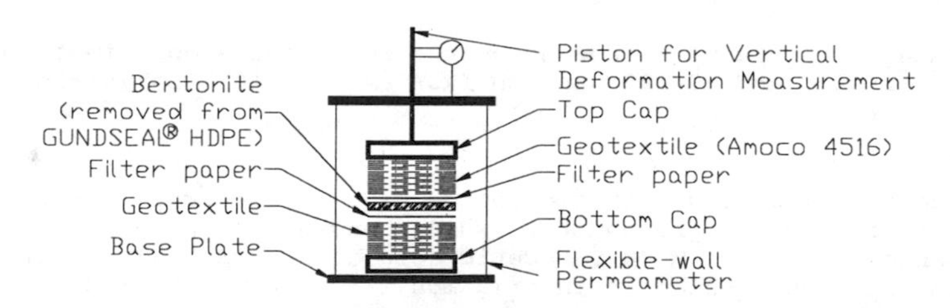

Figure 1 Flexible-Wall Permeameter

CHEMICAL ANALYSIS

Three chemical analyses were performed in this research: permeant liquid analysis, bentonite cation extraction, and bentonite cation exchange capacity. Detailed discussion of these follows.

Permeant Liquid Analysis: Analysis of the permeant liquids were performed on the leached samples to determine concentrations of Na^+, K^+, Mg^{2+}, Ca^{2+} ions, electrical conductance, and pH. The pH of the leachate was measured at the time the samples underwent chemical analysis. Analyses of the permeant liquid for cations consisted of using both Inductively Coupled Plasma (ICP) in accordance with EPA Method 200.7 and a Flame Atomic Absorption Spectrophotometer using EPA Methods 215.1 for calcium, 242.1 for magnesium, 258.1 for potassium, and 273.1

for sodium, respectively.

Cation Extraction: Chemical analysis of the three bentonite clays (BARA-KADE® LD-16, ENVIRO-SEAL® LD-16, and GUNDSEAL®) was performed to determine the cation removal that could be expected in an acidic environment over a period of time. An aggressive nitric acid was used to flush the ions from the bentonite in order to determine an end-point at which removal of the cations would be expected from long-term exposure to the permeant liquids. The laboratory procedure consisted of leaching 0.50 gram of solid bentonite in 50 ml of 10% HNO_3, then filtering and analyzing the liquid for the referenced cations using Flame Atomic Absorption. Results of this test are shown in Table 2, in which the values of the removed cation concentrations for the nitric acid extraction are presented for the three bentonites tested.

Table 2 Chemical Assay of Bentonite Materials

NITRIC ACID EXTRACTION					
MASS SAMPLE 0.50g		Initial	Initial	Initial	Initial
		mg/l	mg/l	mg/l	mg/l
		Ca^{2+}	Mg^{2+}	Na^{+}	K^{+}
ENVIRO-SEAL® LD-16		57.20	16.30	160.00	7.08
BARA-KADE® LD-16		90.10	16.40	150.00	7.82
GUNDSEAL®		**71.20**	**12.70**	**140.00**	**7.15**
FINAL CATION CONCENTRATIONS OF GUNDSEAL® BENTONITE AFTER PERMEATION					
		Final	Final	Final	Final
		mg/l	mg/l	mg/l	mg/l
		Ca^{2+}	Mg^{2+}	Na^{+}	K'
Bentonite	CELL 1	152.00	16.10	1.00	2.00
Bentonite	CELL 2	68.00	14.28	20.69	5.00
Permeant	CELL 1	49200.00	0.00	0.00	0.00
Permeant	CELL 2	0.00	0.00	0.00	0.00
PREDICTED MASS OF GUNDSEAL® BENTONITE IONS					
MASS SAMPLE (g)		Initial	Initial	Initial	Initial
		Ca^{2+} (mg)	Mg^{2+} (mg)	Na^{+} (mg)	K^{+} (mg)
18.02[a]	CELL 1	128.30	22.89	252.28	12.88
16.78[a]	CELL 2	119.47	21.31	234.92	12.00
		Final	Final	Final	Final
18.02[a]	CELL 1	273.90	29.01	1.80	3.60
16.78[a]	CELL 2	114.10	23.96	34.72	8.39

a= mass values include moisture content of 17.7%

The sodium values from the nitric acid extraction were used as the baseline for the predicted mass computations with the GUNDSEAL® bentonite permeated in the permeameters. The final cation concentrations for the GUNDSEAL® bentonite after exposure in the permeameter cells are also presented in Table 2.

The predicted cation mass for the GUNDSEAL® bentonite shown in Table 2 is calculated from the values of the ratio of cations from the actual bentonite GCL samples using the ratio of cations on the 0.5g samples leached by the nitric acid extraction. The values are for comparison with the ion concentrations obtained using the acetic acid and acetic acid / calcium acetate permeant liquids. This information shows the ion removal differences obtained when attempting to predict ion concentrations using a surrogate permeant compared with the actual design permeant.

Cation Exchange Capacity: Cation Exchange Capacities (CEC) were measured for the bentonite portion of GUNDSEAL®. This testing included determining the soluble salts, exchangeable cations, exchangeable anions, exchangeable acidity, electrical conductance, and pH. The results are tabulated in Table 3.

Table 3 Chemical Properties of GUNDSEAL® Bentonite

Property	Value[a]
Electrical Conductance at 25°C	704 μS/cm (avg. four values)
pH at 25°C	7.57 (avg. four values)
SOLUBLE SALTS	mg/l
Fe^{3+}	6.71
Al^{3+}	27.87
Ca^{2+}	1.54
Mg^{2+}	4.54
Na^{+}	18.83
K^{+}	0.91
Cl^{-}	31
SO_4^{-}	4.0
EXCHANGEABLE CATIONS	mg/l (meq/100g)
Ca^{2+}	61.85 (17.18)
Mg^{2+}	20.94 (10.07)
Fe^{3+}	0.01 (0.01)
Al^{3+}	0.03 (0.03)
Mn^{2+}	0.03 (0.01)
Na^{+}	78.0 (19.49)
K^{+}	12.97 (1.94)
Exchangeable Acidity	(6.13)
EXCHANGE CAPACITIES	meq/100g
Cation Exchange Capacity, CEC	55
Anion Exchange Capacity, AEC	< 1

a - average of three measurements

The procedure for determination of the soluble salts ratio was Rhoades (6). The soil/water ratio was 1:5 and no deviations from the original procedures were made. The exchangeable acidity procedure followed was Rhoades (6), Chapter 10 EXCHANGEABLE ACIDITY - $BaCl_2$ - TEA EXTRACTION. In this procedure 2.5 g of soil was used with a 10 ml buffer solution, 40 ml replacement solution and titration with 0.1 N HCL. The pH procedure followed Rhoades (6) where the 1:1 ratio of soil to extract used 10g soil to 10 ml deionized - distilled water. The extract was subsequently used for the electrical conductivity (EC) measurement.

The exchangeable anions, exchangeable cations, CEC and AEC were performed following the procedure by Hendershot and Duquette (1986) as referenced by Rader (7). This method uses barium chloride ($BaCl_2$) as a rapid means of determining the cation exchange capacity and exchangeable cations. This procedure was used because of its applicability to environmental soil problems where the CEC at the pH of the GCL in the field was of relative importance.

RESULTS & DISCUSSIONS

pH

Graphs of pH vs pore volumes of flow for the two permeability tests are shown in Figures 2(a) and 3(a). For both cells, the pH was not immediately determined in the effluent containers but was performed when the samples were analyzed for ion concentrations. Care was taken to seal the samples and minimize the release of aqueous phase carbon dioxide. The pH was not measured in either of the permeant supply cells during the bentonite permeation. Delaying the pH measurement may have resulted in increases in pH with time, however to what extent this may have occurred is unknown.

In Figure 2(a), the pH shows an immediate increase from 3.2 to 3.4, indicating a loss in protons in the permeant effluent. The permeant then gains protons after approximately 10 pore volumes of flow and decreases to a pH of 2.7. This effect in pH may be attributed to buffering of the permeant by the bentonite.

Table 3 lists an average pH measurement of four GUNDSEAL® bentonite samples at 7.57 with a standard deviation of 0.25. This value is lower than data presented by Shackelford (8) where a pH of 8.7 is reported for processed bentonite, and a pH of 9.8 reported by the BENTONITE Corporation (2).

HYDRAULIC CONDUCTIVITY

Measurements of the hydraulic conductivity for the two test cells were performed and are shown in Figures 2(b) and 3(b). Determination of the pore volume quantity for the swellable clay was based on the initial soil properties.

For Cell #1, Figure 2(b), the hydraulic conductivity continues to increase from an initial value of 4.2×10^{-9} cm/s after one pore volume to 4.3×10^{-6} cm/s after permeation of 78 pore volumes (1720 hrs). For Cell #2 Figure 3(b), the hydraulic conductivity decreased from an initial value of 2×10^{-8} cm/s after one pore volume to a low value of 1.1×10^{-9} cm/s after eight pore volumes (1784 hrs), and increased to 3.3×10^{-9} cm/s after ten pore volumes (1900 hrs).

Daniel and Ruhl (5) reported a hydraulic conductivity of 2×10^{-5} cm/s for a non-prehydrated GUNDSEAL® bentonite sample when exposed to a simulated municipal solid waste leachate that contained a 0.15M acetic acid solution (pH=4.35) and dissolved calcium ranging from 360 to 1,000 mg/l. These results are comparable to those obtained in Cell #1.

Daniel and Ruhl (5) also reported a hydraulic conductivity of 2×10^{-7} cm/s for a non-prehydrated GUNDSEAL® sample when exposed to a strong acid, 0.1M HCL having a pH=1. Possible reasons for the differences between Daniel and Ruhl's results and those for Cell #2 are: 1) chemical reactions within the bentonite were not completed, 2) insufficient time for the hydraulic conductivity to reach steady conditions, and 3) the effective stress at the tail water end may have differed from Daniel and Ruhl's.

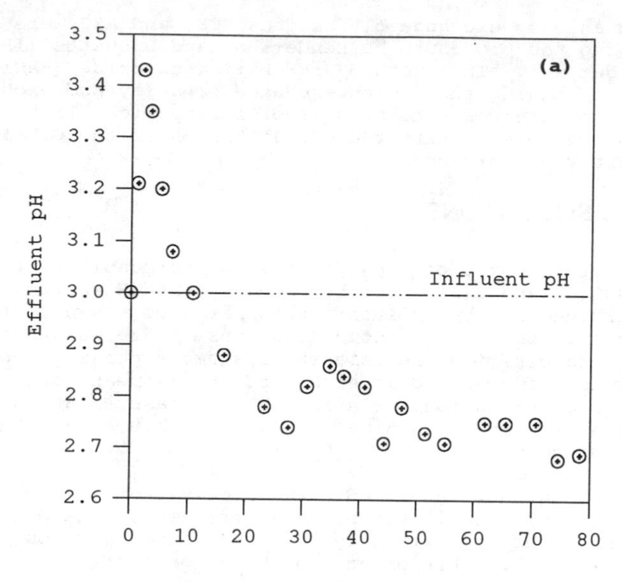

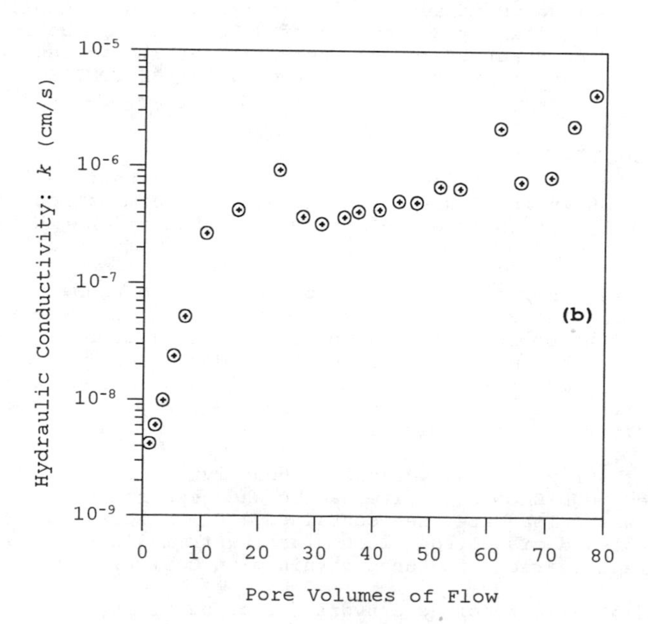

Figure 2 Cell #1 (0.2M $CaCO_3$ + 0.03M CH_3COOH)
(a) pH vs Pore Volumes of Flow
(b) Hydraulic Conductivity vs Pore Volumes of Flow

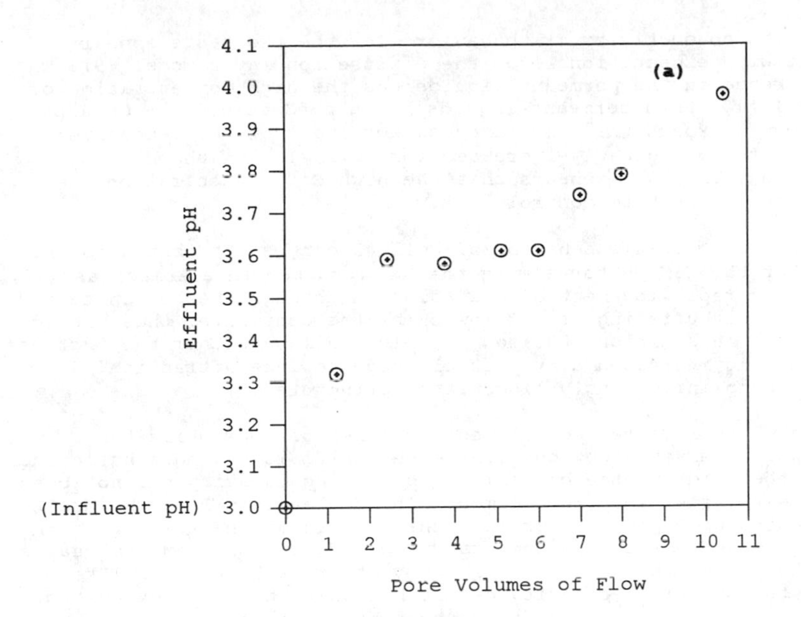

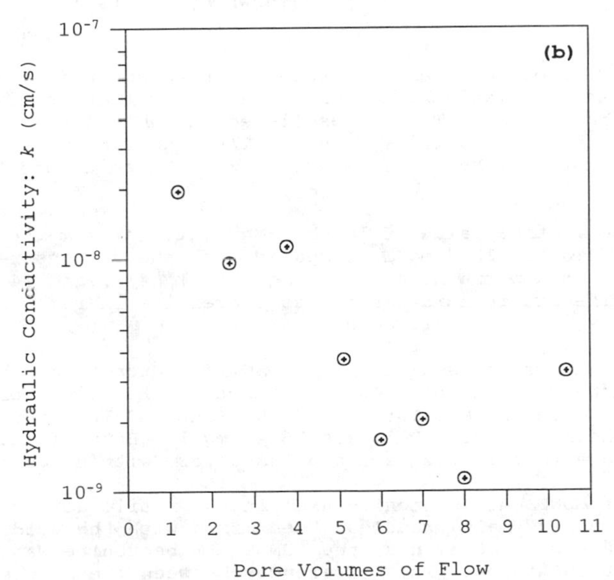

Figure 3 Cell #2 (0.03M CH_3COOH)
(a) pH vs Pore Volumes of Flow
(b) Hydraulic Conductivity vs Pore Volumes of Flow

The hydraulic conductivity (*k*) behaviors for the two tests appear consistent with expectations from the diffuse ion layer model (9). The only difference in the permeant liquids was the high concentration of Ca^{2+} in Cell #1. Both permeant liquids had a pH=3 at 0.03M. On a pH basis we would expect the same response for the *k* of the bentonite; however, we found *k* greatly increased for Cell #1 with the high Ca^{2+} concentration. Thus, it appears that the high Ca^{2+} concentration dominates the short-term control of *k*.

The *k* - pH - pore volume behavior of Cell #1 can be attributed to the process that the Ca^{2+} exchange with the Na^{+} resulted in a *k* increase(10, 11), allowing rapid movement of the solution through the bentonite and little time for buffering of the low pH by the bentonite, thus the low pH effluent. In addition, the low pH solution may further the increase in *k* by causing increased clay particle edge to face attraction resulting in an increasingly flocculated structure (9).

For Cell #2, the *k* decreased with pore volumes of flow while the effluent pH increased. Obviously, the bentonite is having a buffering effect on the acid and the bentonite's buffering capacity has not been exhausted even after 10 pore volumes. The decrease in *k* may be due to a dissolutioning-precipitation process that occurs at the acid front in the bentonite and results in temporary clogging of pores and decreases in *k*. This process has been described by others (12). It is likely that when the buffering capacity of the bentonite has been exhausted, the *k* will be greatly increased, as shown by others (13).

ION EXCHANGE

The quantity of cations removed based on the predicted values of available cations, Table 2, versus the actual values of leached cations are listed in Table 4. The values listed in Table 4 are the difference between the initial and final values. The negative values for the predicted removal of Mg^{2+} cations may be attributed to the following reasons:

a) The initial value of leached Mg^{2+} cations from the nitric acid extraction is 12.7 mg/l, compared with the cations removed by the acetic acid permeant are 16.1 mg/l (Cell #1), and 14.3 mg/l (Cell #2). The nitric acid was not as aggressive in removing the Mg^{2+} cations as the acetic acid.

b) The adhesive used in the GUNDSEAL® bentonite may have reacted with the Mg^{2+} cations. This is based on the levels of Mg^{2+} cations reported in Table 2 for the ENVIRO-SEAL® LD-16 are 16.30 mg/l and for the BARA-KADE® LD-16 are 16.40 mg/l. Both of these specimens were analyzed in the raw, processed form without adhesive.

c) The GUNDSEAL® bentonite used in the nitric acid extraction was prepared in small granules, then mixed with the acid. For the permeameter experiments, the GUNDSEAL® bentonite was tested in an intact condition. The differences between the sample testing conditions are: 1) for the nitric acid, more surface area of the granulated bentonite would be exposed for chemical contact, and 2) the permeameter testing resulted in a longer exposure time as compared with the nitric acid which was a wash process.

Table 4 Comparisons of Predicted versus Actual Cation Removal

Cell		Na^+ (mg)	K^+ (mg)	Mg^{2+} (mg)
#1: Acetic Acid + Calcium	Predicted Actual	250.48 185.33	9.28 6.99	-6.12 29.08
#2: Acetic Acid	Predicted Actual	200.20 82.94	3.61 2.46	-2.65 4.00

Results of the CEC testing on four GUNDSEAL® bentonite samples averaged 55 meq/100g. This average is less than that expected for sodium bentonite which is typically in the range of 80 to 150 (9). Possible reasons for differences between the experimental and published values are: a) experimental procedures followed, and b) the adhesive binding the bentonite to the geomembrane on the GUNDSEAL® samples may have chemically altered the clay.

Graphs of the Ca^{2+}, Mg^{2+}, Na^+, and K^+ ion concentrations (mg/l) analyzed in the leached permeant fluid versus pore volume of flow for Cells #1 and #2 are shown in Figures 4(a) and 5(a). Figures 4(b) and 5(b) show the cumulative ions removed from the bentonite specimens in the permeameters versus pore volume of flow occurring during the permeation period. The cumulative ions are represented in (mg) and were determined using the mass balance data presented in Table 2.

SODIUM ION EXCHANGE: For cell #1 (Figure #4(a)), the GCL specimen continues to leach Na^+ from a high value of 1200 mg/l after 2 pore volumes of flow to 5.1 mg/l after 78 pore volumes (pH = 2.69). Concentration of Ca^{2+} in the effluent increases after approximately 30 pore volumes of flow. The nitric acid extraction from Table 2 shows a final Na^+ concentration in the removed specimen at 1 mg/l and the final concentration of Na^+ leached from the acetic acid at 5.1 mg/l.

Figure 4(a)shows that concentrations of Mg^{2+}, Na^+, and K^+ continue to be removed from the bentonite in small amounts after 78 pore volumes of flow indicating that the reactions have not reached equilibrium. This observation is similar to the decreasing pH shown in Figure 2(a), where the magnitude of released protons from the bentonite going into solution is consistent with the magnitude of the leached Mg^{2+}, Na^+, and K^+ ions in the effluent.

The pH drop from 3.0 to 2.7 after 10 pore volumes corresponds to a 2×10^{-3} M detected increase in Mg^{2+}, Na^+, and K^+ ions in the effluent, or 1×10^{-3} M increase in H^+ protons. This increase is approximately equal to the reduction in Mg^{2+}, Na^+, and K^+ ions from the bentonite. This is evidenced where the 10 mg/l reduction in Mg^{2+}, taking place after the initial 10 pore volumes of flow, reduces to 0.4×10^{-3} M in removed ions from the bentonite. Which all sums up to the point that the continued introduction of Ca^{2+} may result in a release of protons from the bentonite causing the effluent to increase in acidity.

Figure #4(b) shows the cumulative ions removed from the bentonite. This figure does not clearly reflect the slight increase in removed ions from the bentonite occurring after approximately 20 pore volumes of flow. For cell #2, Figure#5(a), the incremental values of leached Na^+ in the acetic acid permeant liquid remain steady at approximately 800 mg/l for almost 1900 hrs (10 pore volumes). Figure #5(b) shows a steadily

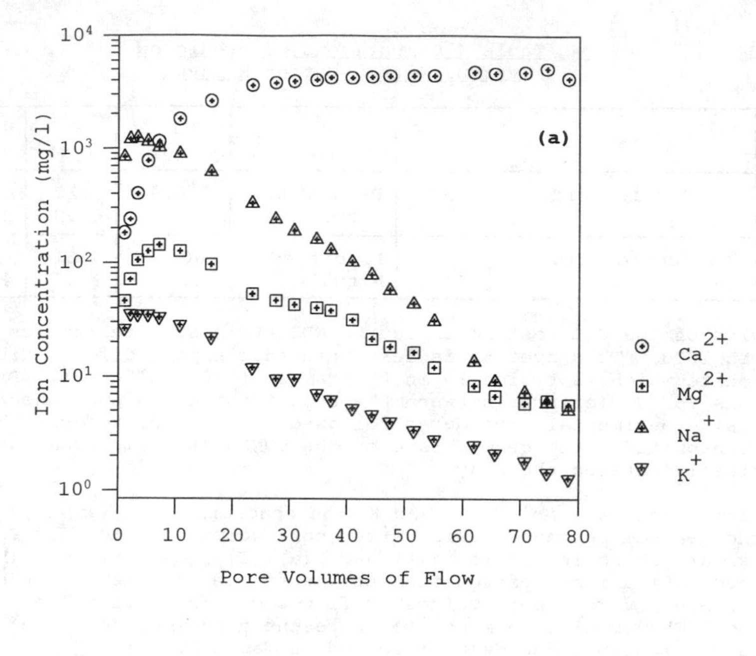

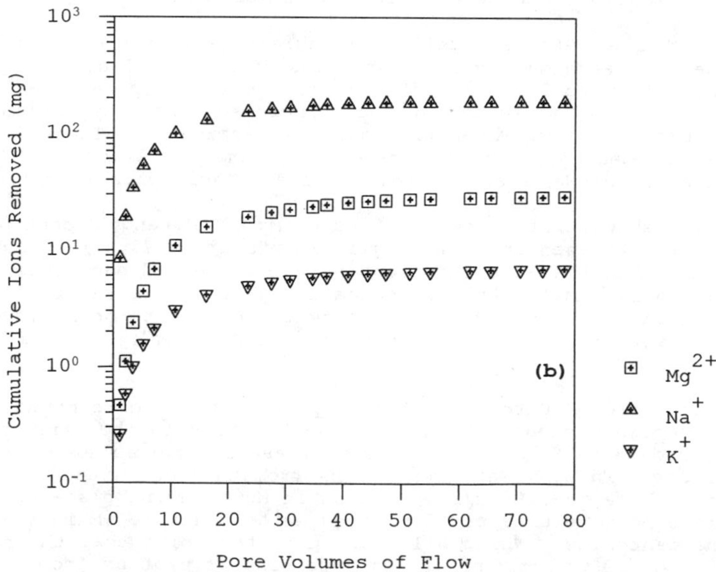

Figure 4 Cell #1 (0.2M $CaCO_3$ + 0.03M CH_3COOH)
(a) Ion Concentration vs Pore Volumes of Flow
(b) Cumulative Ions Removed vs Pore Volumes of Flow

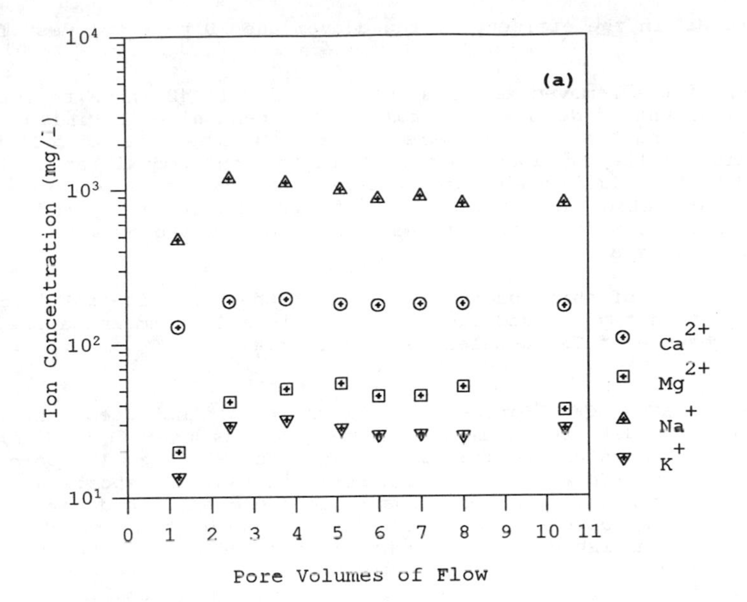

Cumulative Ions Removed (mg)
10^2
10^1
10^0
10^{-1}
(b)
0 1 2 3 4 5 6 7 8 9 10 11
Pore Volumes of Flow
Ca^{2+}
Mg^{2+}
Na^{+}
K^{+}

Figure 5 Cell #2 (0.03M CH_3COOH)
(a) Ion Concentration vs Pore Volumes of Flow
(b) Cumulative Ions Removed vs Pore Volumes of Flow

increasing Na^+ in the effluent averaged over the 10 pore volumes of flow.

Comparison of the removed Na^+ from Cell #1 and Cell #2 shows that the cumulative amount of Na^+ leached from the GCL bentonite is similar for both tests during the first 10 pore volumes. In the case of Cell #1, the addition of the Ca^{2+} ions increased the Na^+ ion removal rate. The maximum detected limit was reached after 1000 hrs (3 pore volumes) then the Na^+ concentration detected in the effluent significantly reduced. For Cell #2 the concentration of the removed Na^+ is steady after 1750 hrs (8 pore volumes).

The significance of this observation is that the bentonite's Na^+ ions may be significantly removed due to the acetic acid permeant alone, and that the presence of Ca^{2+} accelerates the removal.

POTASSIUM ION EXCHANGE: For Cell #1, Figure #4(a) indicates that the concentration of detected K^+ in the effluent was reduced from 33 mg/l at 1264 hrs, (7 pore volumes), and to 1.28 mg/l at 1720 hrs, (78 pore volumes). The nitric acid extraction shown in Table 2 reports a concentration of 2.0 mg/l after exposure to the acetic acid enriched with calcium. The detection of K^+ in the effluent of Cell #1 significantly diminished after 1200 hrs (4 pore volumes). For the acetic acid permeant in Cell #2, Figure 5(a), the K^+ shows a steady level of 25 mg/l at approximately 1900 hrs (10 pore volumes).

MAGNESIUM ION EXCHANGE: For Cell #1, Figure 4(a) shows the Mg^{2+} to be significantly reduced after 1300 hrs (16 pore volumes). In Cell #2, Figure 5(a), the Mg^{2+} appears steady at approximately 50 mg/l.

CALCIUM ION EXCHANGE: For Cell #1, Figure 4(a) shows the Ca^{2+} effluent attaining a concentration of 4,000 mg/l after 20 pore volumes of flow. The level of Ca^{2+} ions detected in the effluent increase continually but does not reach its breakthrough concentration of 4,920 mg/l. The Ca^{2+} continues to cause chemical reactions within the bentonite, which appears as changes in the effluent pH (Figure 2(a)). For Cell #2, Figure 5(a), the concentration of leached Ca^{2+} in the effluent is approximately 180 mg/l throughout the 1900 hrs (10 pore volumes) of exposure.

ELECTRICAL CONDUCTANCE

The Electrical Conductance (EC) for the GUNDSEAL® bentonite listed in Table 3 is 704 μS/cm (microSiemens). This value is more than double the value reported by Shackleford (8) at 300 μS/cm. The EC of the permeants reported in Table 1 are: 6,270 μS/cm (acetic acid and calcium), and 286 μS/cm (acetic acid only).

Graphs of the Electrical Conductance versus Net Pore Volumes of Flow for Cell #1 and Cell #2 are shown in Figure 6. For Cell #1 Figure 6(a), the EC remains reasonably steady at approximately 6,000 μS which is close to that of the permeant liquid. The EC for Cell #2, Figure 6(b), shows consistency with the last three pore volumes being within a range of 3630 to 3230 μS/cm. The significance of this observation is that for the acetic acid, the concentration of ions capable of carrying an electrical current increased to approximately half that of the calcium enriched permeant liquid in Cell #1; and that the bentonite was the source of these ions.

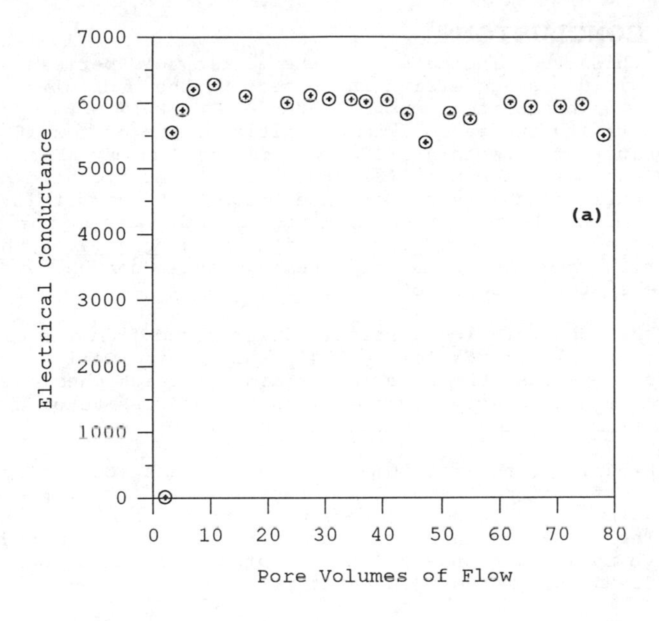

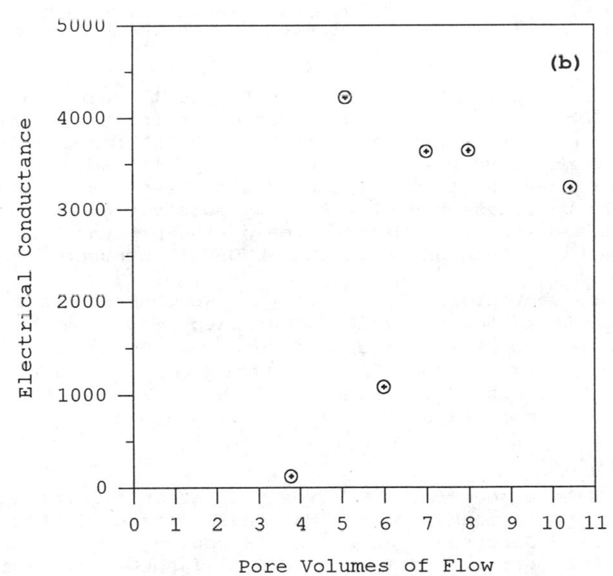

Figure 6 Electrical Conductance vs Pore Volumes of Flow
(a) Cell #1 (0.2M $CaCO_3$ + 0.03M CH_3COOH)
(b) Cell #2 (0.03M CH_3COOH)

SUMMARY & CONCLUSIONS

The results of this study indicate that the acetic acid permeant enriched with calcium is aggressive in increasing the hydraulic conductivity of the bentonite component of the GUNDSEAL® GCL. The hydraulic conductivity increased from an initial value of 4×10^{-9} cm/s after 1 pore volume of flow to 4×10^{-6} cm/s after 78 pore volumes of flow. For the case of the bentonite exposed to the acetic acid permeant alone, the hydraulic conductivity decreased from 2×10^{-8} cm/s to a low of 3.3×10^{-9} cm/s. In the latter case, chemical equilibrium may not have been achieved as noted by the pH of the effluent being greater than that for the influent. Thus, the final reported hydraulic conductivity is only applicable at the 10 pore volumes of flow.

In both cases the total predicted sodium ions removed from the clay was higher than the actual by: 26% for Cell #1, and 58% for Cell #2. Therefore we conclude that the experiments did not reach chemical equilibrium or the nitric acid may have been too aggressive of a solution for determining the ultimate ion removal.

The bentonite exposed to the calcium-enriched permeant, Cell #1, resulted in the effluent pH to initially rise from 3.0 to 3.4, then after 10 pore volumes, decrease and remain at approximately 2.7 for 78 pore volumes. The reduction in pH may be attributed to the introduction of the Ca^{2+} which may result in a release of protons from the bentonite causing the effluent to increase in acidity.

The pH of the effluent in Cell #2 increased gradually during the 10 pore volumes of flow. The increase is believed due to the buffering capacity of the bentonite.

Results of the Liquid Limit tests showed a LL=330 for GUNDSEAL® compared with LL=675 and LL=779 for the two unprocessed bentonites. Comparison between the cation exchange capacity for the GUNDSEAL® bentonite showed a value of CEC=55 meq/100g which is lower than published values for bentonite. Differences in the lower Liquid Limit and the CEC between the GUNDSEAL® and the unprocessed bentonites may possibly be attributed to a reduction in the exposed clay surface area by the presence of the adhesive mixed with the bentonite in the GUNDSEAL® product.

The final moisture contents recorded for the two GUNDSEAL® bentonite specimens tested showed for the calcium-enriched permeant, w=172%, and for acetic acid only, w=231%. This difference may be due to the almost complete exchange of Na^{+} for Ca^{2+} from the bentonite exposed to the calcium-enriched permeant liquid compared with the Na^{+} concentration from the bentonite sample exposed to acetic acid alone.

ACKNOWLEDGMENT

The authors would like to express their appreciation for the in-kind support provided by: Raymond Lovett, (National Research Center for Coal and Energy), Richard Erickson (GSE Lining Technology, Inc.), Charles McAughan (BENTONITE Corp.), James D. Anderson, (JANCO Engineering), and Jim Gorman, (West Virginia University Plant and Soil Sciences).

REFERENCES

(1) Koerner, Robert M., (1994). *Designing with Geosynthetics*, 3rd Ed., Prentice-Hall Inc., Englewood Cliffs, New Jersey.

(2) BENTONITE Corporation (1992), Personal Communication.

(3) Baver, L.D., Gardner, W.H., and Gardner, W.R., (1972). *Soil Physics*, 4th Ed., John Wiley & Sons, New York, N.Y., pg 34.

(4) Krauskopf, K.B. (1979). *Introduction to Geochemistry*, 2nd Ed., McGraw-Hill Book Co., Inc., New York, N.Y., pg 87.

(5) Daniel, D.E., and Ruhl, J.L. (1994)."Effects of Leachate On the Hydraulic Conductivity of Bentonite and Contaminant-Resistant Bentonite in GUNDSEAL®". Final Report, University of Texas.

(6) Rhoades, J.D. (1982)., Chapter 10: Soluble Salts. "*Methods of Soil Analysis, Part 2 - Chemical and Microbiological Properties*," 2nd Ed., A.L. Page, R.H. Miller, and D.R. Keeny, eds., American Society of Agronomy, Soil Science Society of America., Madison, WI pp.149-157.

(7) Rader, In Soil Sampling and Methods of Analysis (1993) Canadian Soil Science Society, pp 168-170.

(8) Shackelford, C.D. (1994). "Waste-Soil Interactions that Alter Hydraulic Conductivity, *ASTM STP 1142*," David E. Daniel and Stephen J. Trautwein, eds., ASTM, Philadelphia, P.A., 111-168.

(9) Mitchell, J.K. (1993). *Fundamentals of Soil Behavior*, 2nd Ed., John Wiley and Sons, Inc., New York, N.Y.

(10) Grim, R.E. (1968). *Clay Mineralogy*, 2nd Ed., McGraw-Hill Book Co., Inc., New York, N.Y.

(11) Gleason, M. (1993)., "*Comparison of Calcium and Sodium Smectite Clays for Geotechnical and Environmental Applications*," Thesis, University of Texas - Austin, 200p.

(12) Bowders, J.J. (1988)., "Termination Criteria for Clay Permeability Testing," Discussion of Paper by Pierce and Witter, *Journal of Geotechnical Engineering*, ASCE, 114(8): 947-949.

(13) Bowders, J.J. and Daniel, D.E., (1987) "Hydraulic Conductivity of Compacted Clay to Dilute Organic Chemicals," *Journal of Geotechnical Engineering*, ASCE, 113(12):1432-1448.

Hydraulic Conductivity Testing Issues and Methods

Gérard Didier, [1] Loretta Comeaga[2]

INFLUENCE OF INITIAL HYDRATION CONDITIONS ON GCL LEACHATE PERMEABILITY

REFERENCE: Didier, G. and Comeaga, L., **"Influence of Initial Hydration Conditions on GCL Leachate Permeability,"** *Testing and Acceptance Criteria for Geosynthetic Clay Liners, ASTM STP 1308,* Larry W. Well, Ed., American Society for Testing and Materials, 1997.

ABSTRACT: Modern landfill liners typically contain several geosynthetics and natural components integrated into a system whose primary function is the containment of waste and leachate. One of these hydraulic barrier components is represented by Geosynthetic Clay Liners (GCL). The GCLs are factory manufactured hydraulic barriers consisting of bentonite clay supported by geotextiles or geomembranes. Since the waterproofing function is assured by the bentonite clay it is therefore advisable to verify its chemical compatibility with the leachate and its effect upon the hydraulic conductivity value. According to the climatic situations occurring between the GCLs and waste installation, the leachate exposure moment can correspond to various GCLs hydration degrees. In this context, this paper aims at analyzing the total or partially hydrated GCL's behavior after a long time exposure to leachate. The permeability tests carried out for two needle-punched GCLs in three situations (saturated with water, water partially hydrated and saturated with leachate) show an important variation of the permeability with the hydration degree.

KEYWORDS: Geosynthetic Clay Liner, bentonite, leachate, interaction, swelling, permeability, oedopermeameter

---

[1] Doctor, Geotechnical Department, INSA Lyon, 20, av. A. Einstein, 69621 Villeurbanne Cedex, France

[2] Postgraduate student, Geotechnical Department, INSA Lyon, 20, av. A. Einstein, 69621 Villeurbanne Cedex, France

Nowadays, GCLs are more and more often used as sealing elements in waste landfills. During the last few years, research carried out in order to evaluate and analyze the advantages offered by these materials versus compacted clay and geomembranes has been intensified. Nevertheless, the permeability of the existing GCLs is not analyzed on particular site conditions.

Moreover, different manufacturers and research laboratories use various methods. The aim of this research work is, on the one hand, to provide a unitary way of evaluating different products performances and, on the other hand, to study the influence of the initial moisture content on GCL permeability.

The main reason for carrying out this work lies in the real placement and operation conditions observed on several construction sites. In fact, even if the GCL is placed dry, when its placing and the landfill operation are deferred, the moisture state of the GCL is strongly dependent on the climatic conditions.

Thus, the GCL used as a liner system element or in the cover system may appear in the following detrimental situations:

- GCL completely saturated with water (installation on rainy days or artificially sprinkling),
- GCL partially hydrated (by rainfall or artificially sprinkling),
- GCL placed in dry conditions, in direct contact with the waste.

The apparatus used for carrying out the laboratory tests, built at INSA Lyon, France, Geotechnical Laboratory, is a large diameter oedopermeameter. It allowed us to follow both saturation and swelling kinetics and to measure the permeability.

The tests were conducted on two needle-punched GCLs, containing sodium bentonite, saturated or partially hydrated with water or synthetic leachate.

DESCRIPTION OF THE APPARATUS

The apparatus is a large diameter oedopermeameter with a constant hydraulic head. It allows us to measure both in- and out-flow under a given normal stress.

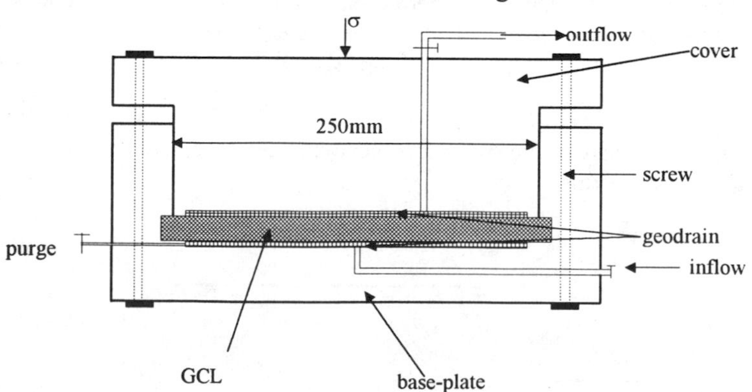

FIG 1-- Apparatus diagram

It is comprised of two basic parts: a base-plate and a cover, each one provided respectively with an intake and an outlet for the percolating fluid (water or leachate). The apparatus is made of polyethylene. The two parts are joined together by 6 screws, which are used only during the permeability measurement (to prevent the lift up of the sample), after the swelling stabilization.

The sample is placed between 2 porous draining elements (geodrains).

The inner cell diameter is 250mm, but the sample size is 3mm larger in order to ensure lateral edge sealing.

The fluid inflows through the bottom and is recovered through the top of the cell.

The normal stress is applied in the same manner that for a classical oedometer, by direct loading.

The border of the dry sample are slightly wetted in order to minimize the bentonite losses. A bentonite seal is used all around the sample during its placement (see detail - figure 2) The bentonite paste is made of a mixture of demineralized water and bentonite recovered at the same location as the GCL sample.

The sample's initial thickness is measured with reference to the distance between the cover and the base plate when the cell is empty and under the desired load.

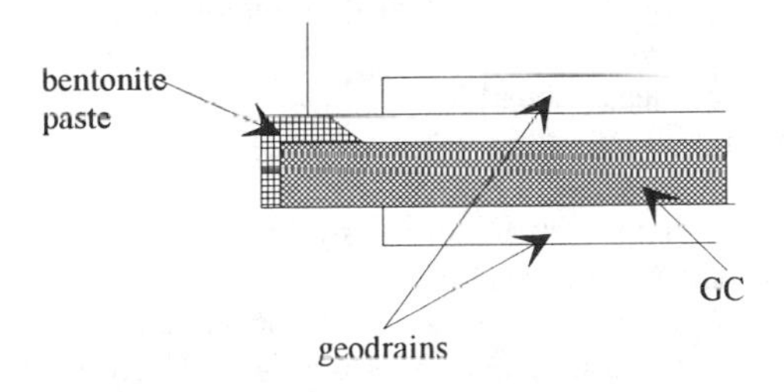

FIG. 2-- Lateral sealing

TESTING PROCEDURE FOR WATER OR LEACHATE SATURATED SAMPLES

The applied procedure for the saturated samples (water or leachate) contains the following steps:

1. initial water content measurement;
2. the dry GCL sample is placed in the cell;
3. sample initial thickness measurement;
4. consolidation stage under a given normal stress;
5. sample thickness measurement under load;
6. continuous saturation (demineralized water or leachate) and monitor of swelling and infiltration kinetics until stabilization. During this stage the hydraulic head is null and the screws are not fixed. The results are interpreted progressively using the hyperbolic law described below (estimation of the swelling for an infinite time, Δh_∞). The saturation stage is complete when the sample swelling has reached at least 90% of the estimate final swelling, Δh_∞.

7. when at least 90% of the final swelling is reached, the piston is immobilized by manual screwing of the nuts, in order to avoid the lift up of the sample during the permeability measurement, if the hydraulic head is higher then the applied normal stress;
8. demineralized water or leachate permeability measurement under the desired hydraulic head;
9. leachate admission (for the water saturated samples);
10. leachate permeability measurement (for the water saturated samples);
11. effluent chemical analysis (by fractions);
12. final water content measurement.

TESTING PROCEDURE FOR PARTIALLY HYDRATED SAMPLES

The applied procedure for the partially hydrated samples was the follow:

1. the GCL samples are hydrated by immersion in demineralized water for a few seconds or minutes;
2. the samples are kept for a week in order to homogenize its wetness over its surface;
3. initial water content measurement;
4. GCL sample placing in the cell;
5. initial thickness measurement;
6. consolidation stage under a given normal stress;
7. thickness measurement under load;
8. the piston is immobilized by manual screwing of the nuts;
9. leachate permeability measurement under the desired hydraulic head;
10. final water content measurement;
11. effluent chemical analysis (by fractions).

The calculations were made using the hyperbolic law for swelling [1]:

$$\Delta h = \Delta h_{\infty} \times \frac{t}{t_{50} + t}, \qquad (1)$$

where:

Δh is the sample swelling (mm),
t is the time (min),
Δh_{∞} is the swelling for an infinite time(mm),
t_{50} is the time corresponding to 50% of the final swelling (min).

This law corresponds to a straight line in the $t/\Delta h$ - t plan for time long enough:

$$t/\Delta h = at + b, \qquad (2)$$

where: $a = 1/\Delta h_{\infty}$ and $b/a = t_{50}$.

For the permeability calculations, the Darcy's law was used.

MATERIALS DESCRIPTIONS

1. The GCLs

The two studied GCLs have the following characteristics:

GCL1 - contains a granular natural sodium bentonite (5 kg/m^2) between two geotextiles, one woven and the other one nonwoven. The bentonite and the geotextiles are held together by needle-punching.

GCL2 - contains a natural sodium bentonite powder (5 kg/m^2) between a nonwoven geotextile and a woven slit film. It is a needle-punched GCL . The sodium bentonite is also encapsulated in the pores of the nonwoven cover layer.

2. The leachate

The leachate used for the tests is a synthetic one, similar to a real municipal waste leachate. It is formulated based on the literature data concerning the chemical composition of the waste leachate (Table 1).

TABLE 1--General chemical composition of the waste leachate [2]

Parameters	Units	Minimum values	Maximum values	Mean geometrical value	Sample number
NH_4^+	mgN/l	0.9	2154	147	76
NO_3^-	mgN/l	0	85	1.2	74
P total	mgP/l	0.1	14	1.55	25
pH		4.9	8.9	6.9	105
Alkalinity	mg $CaCO_3$/l	80	26000	964	45
Conductivity	µS/cm	295	38000	6303	70
Ca^{2+}	mg/l	50	3650	253	57
Mg^{2+}	mg/l	0.6	526	77	57
K^+	mg/l	20	1600	228	51
Na^+	mg/l	35	9500	424	44
Cl^-	mg/l	7	8800	523	79
SO_4^{2-}	mg/l	3	3239	121	52
CH_3COOH	mg/l	1	17000	154	40

In order to have the most pessimistic of the working conditions, Ca^{2+}, Mg^{2+}, NH_4^+, K^+, CH_3COOH, Cl^- and SO_4^{2-} concentrations of more than the mean value and a sodium concentration of about the mean value have been chosen (Table 2).

TABLE 2--Synthetic leachate chemical composition

component		cations		anions	
formula	concentration (g/l)	symbol	concentration (g/l)	symbol	concentration (g/l)
CH_3COOH (acetic acid)	1.2	H^+		CH_3COO^-	1.18
CH_3COONa (sodium acetate)	1.64	Na^+	0.46	CH_3COO^-	1.18
$CaCl_2$ (calcium chloride)	3.323	Ca^{2+}	1.2	Cl^-	2.123
$MgCl_2$ (magnesium chloride)	1.186	Mg^{2+}	0.303	Cl^-	0.883
$MgSO_4$ (magnesium sulfate)	0.481	Mg^{2+}	0.097	SO_4^{2-}	0.384
NH_4Cl (ammonium chloride)	1.483	NH_4^+	0.5	Cl^-	0.983
KOH (potassium hydroxide)	1.119	K^+	0.78	OH^-	0.319

The Ca/Na ratio is also very important. Of all the possible compounds, the ones having the most important influence on both swelling and permeability have been chosen. As it is known that the bentonite permeability increases generally as a response to saline solution action, these chemical compounds are very well represented. The Na, Ca, Mg and K cations have a particular interest due to their great influence on the bentonite swelling.

Among the organic substances, the acetic acid and the sodium acetate have been chosen. It has been noticed that the alkaline salts of the short chain fatty acids, acetic acid in particular, can get into the interlayer space and dissociate the bentonite layer [3]. Compounds having a low water solubility, e.g. hydrocarbons, do not have a significant influence on the bentonite permeability and they are not represented in this leachate. The same for the water soluble organic substances (alcohol, ketones) who's effect is important only at high concentrations. As the clays are permeable for non-polar compounds only in pure form, which are hardly ever found in practice, such substances have not been taken into account for the leachate composition.

The leachate has a pH = 5.40 and an electrical conductivity of 1.16 S/m.

3. GCL's bentonite characterization

For a summary description of the bentonites, the free swelling test has been chosen, which is a simple and effective method of checking the swelling capacity of bentonites. The quality of a sodium-bentonite can, in particular, be determined by it's high swelling volume and a lengthy period of swelling [4]

The tests were carried out with demineralized water and leachate (table 3).

TABLE 3--Free swelling test results (for 2g dry clay into 100ml solution, after 48 hours)

GCL	V (ml) with demineralized water	V (ml) with leachate
1	29	11
2	29	10

RESULTS

1. Water saturated samples

The tests were conducted under a normal stress of 8kN/m^2, which corresponds to a 0.50m thick confining layer, usually used for liner systems in municipal waste storage.

The swelling and infiltration kinetics are presented in figures 3 and 4.

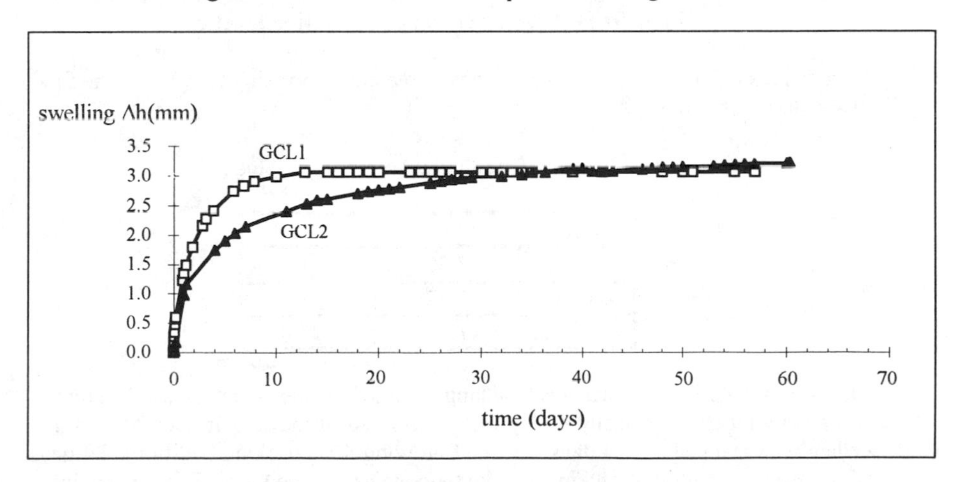

FIG 3-- Demineralized water swelling kinetics

The swelling kinetic obeys the hyperbolic law given in equation (1). Similar, for the infiltration kinetic:

$$V = V_{\infty} \frac{t}{c+t}, \tag{3}$$

where:

V is the water infiltrated volume (cm^3),
t is the time (min),
V_{∞} is the final swelling corresponding volume (cm^3),
c is the time corresponding to a x% of the final volume (min).

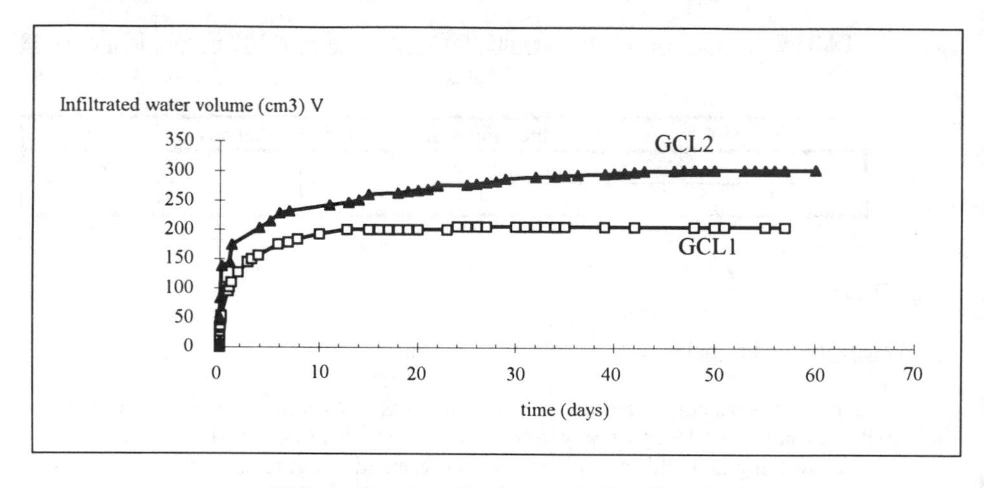

FIG. 4-- Demineralized water infiltration kinetics

The values of the final swelling and of the time corresponding to 50% of the final swelling are shown in table 4.

TABLE 4-- Swelling characteristic values

GCL	1	2
Δh_{∞} (mm)	3.39	3.45
t_{50} (days)	1.28	4.89
t_{90} (days)	12.8	39.04

The two GCLs have similar final swelling values, but their kinetics are different. GCL2 has a slow kinetic, especially during the second half of the test. In fact, 50% of its final swelling was reached after 4 days, but the following 40% took an additional 34 days.

Concerning the infiltration kinetics, the tendencies are the same. Even if the final swelling is very close for GCL1 and GCL2, the infiltrated water quantities are very different.

Figure 5 shows the demineralized water permeability values corresponding to a hydraulic head of 0.80m.

The permeability values vary between 2×10^{-11}m/s for GCL2 and 3×10^{-11}m/s for the GCL1.

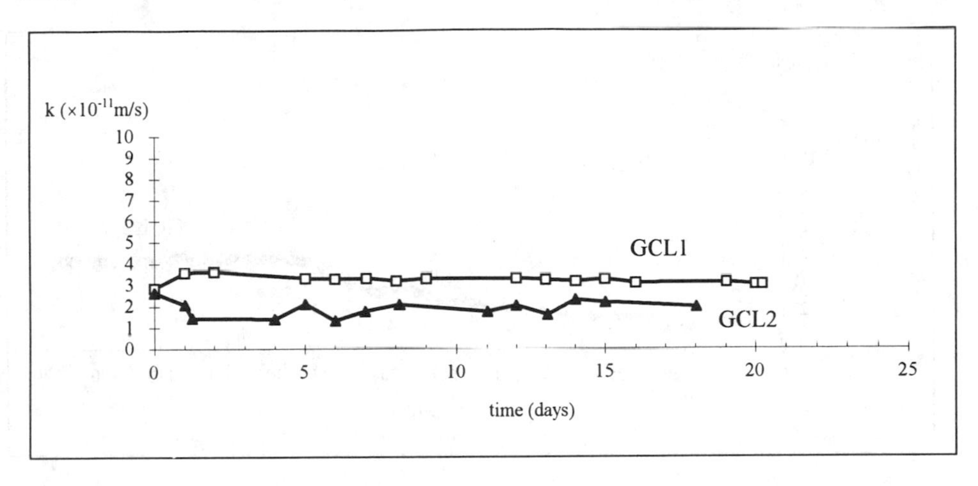

FIG. 5-- GCL permeability versus time evolution for permeation with demineralized water (hydraulic head 0.80m, normal stress 8kN/m^2)

In figure 6 is shown the obtained values for the leachate permeability of GCL1 under a hydraulic head of 0.80m.

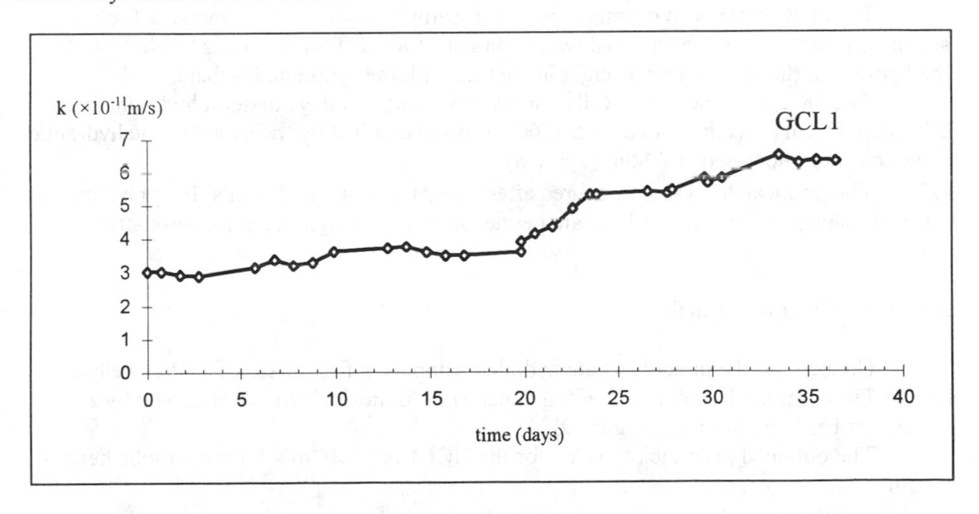

FIG. 6--Leachate permeability for GCL1 under a 0.80m hydraulic head and a 8kN/m^2 normal stress.

The leachate permeability value is 6×10^{-11}m/s.

It is important to note that 37 days permeation in the laboratory under a hydraulic gradient of 75 is equivalent to a 7.7 years permeation under a unit hydraulic gradient.

Figure 7 shows the leachate permeability evolution in time for the GCL2. The permeability value is 4.5×10^{-11}m/s.

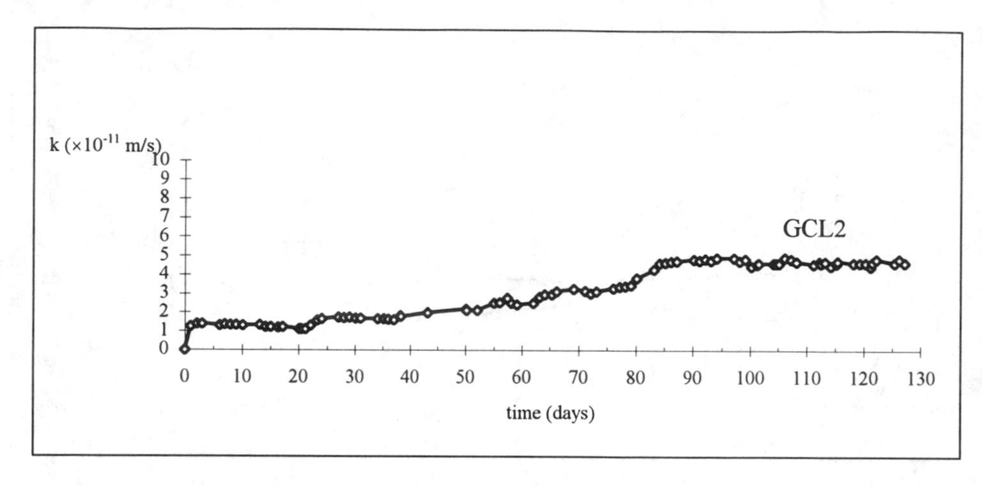

FIG. 7--Leachate permeability for GCL2 under a 0.80m hydraulic head and a $8kN/m^2$ normal stress

2. Partially hydrated samples

Figure 8 presents an example for a GCL sample hydrated by immersion for 10 seconds, which resulted in an initial water content of 80%. This sample is labeled 1a. After the hydration, the sample was brought in contact with the synthetic leachate.

The GCL remained practically impervious during 15 days under a hydraulic head of 0.05m. During this time, a volume of $60cm^3$ was adsorbed by the sample. The hydraulic head was later increased at 0.80m (figure 8).

The permeability value measured after 120 days was 5.5×10^{-9}m/s. It can be noticed a slow evolution in time of the leachate permeability with a significant increase after 70 days.

3. Leachate hydrated samples

Concerning the direct leachate hydrated samples, a first phase of slight swelling (e.g. 0.78mm for GCL1 after 4 days and under $8kN/m^2$ normal stress), followed by a settlement has been observed (figure 9).

The obtained permeability value for the GCL1 is 5×10^{-8}m/s for a hydraulic head of 0.50m.

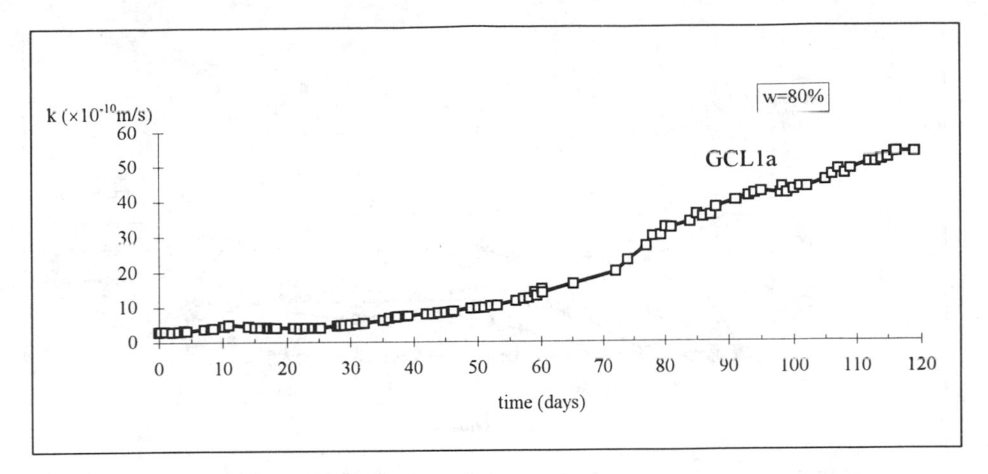

FIG. 8--GCL1a leachate permeability under a 0.80m hydraulic head and a $8kN/m^2$ normal stress.

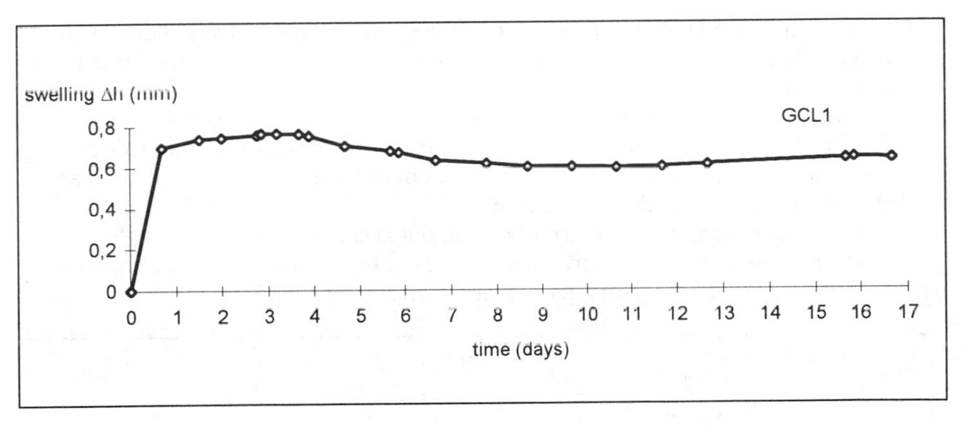

FIG.9--Leachate swelling kinetic for GCL1, under $8kN/m^2$ normal stress

EFFLUENT CHEMICAL ANALYSIS

The results of the chemical analyses carried out for GCL1 and 1a during the permeation with the synthetic leachate are presented.

The pore fraction is the ratio of the effluent volume to the pore volume.

Figure 10 shows the evolution of the concentration in main cations of the effluent versus the pore fraction for the GCL1, under 2.0m hydraulic head.

It can be observed a sodium desorption which is simultaneous with a calcium and potassium adsorption. There is a ionic competition between the Na^+ and K^+ cations, which are producing a transformation of the Na-bentonite into a polycation-bentonite (Ca, Na, Mg).

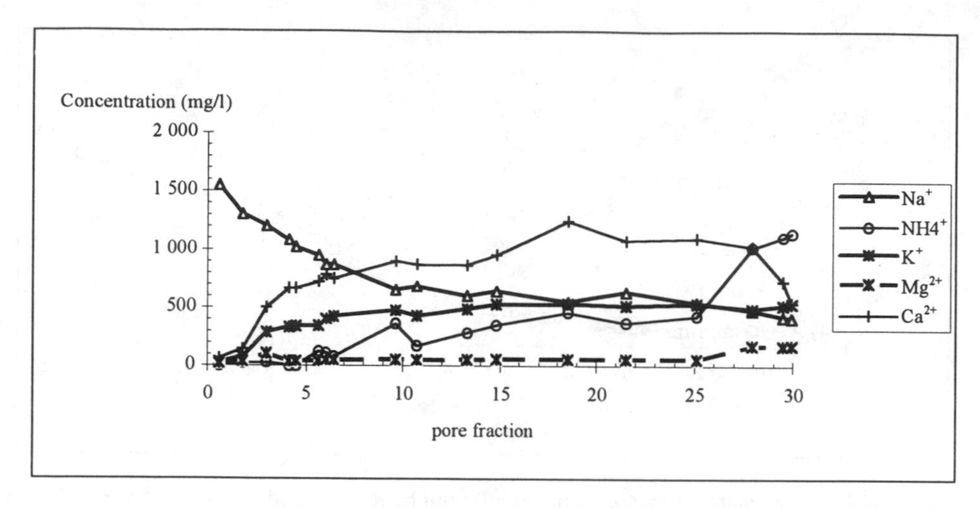

FIG. 10--Effluent cation concentration versus pore fraction for GCL1.

After an effluent volume of about 8 times the pore volume (corresponding with 60 days percolating), the sodium concentration (desorption) becomes equal with the calcium one (adsorption).

After the breakthrough of a volume 8 times the void volume, the lost sodium quantity becomes equal to the adsorbed calcium quantity. A decrease in adsorbed calcium, together with a slight increase in potassium can be noticed after the breakthrough of a leachate volume of 28 times the void volume.

The NH_4^+ and the Mg^{2+} are also adsorbed, but at smaller rates.

The pH and electric conductivity curves show a relative stability of these values, while the chemical element concentration is more variable (figure 11).

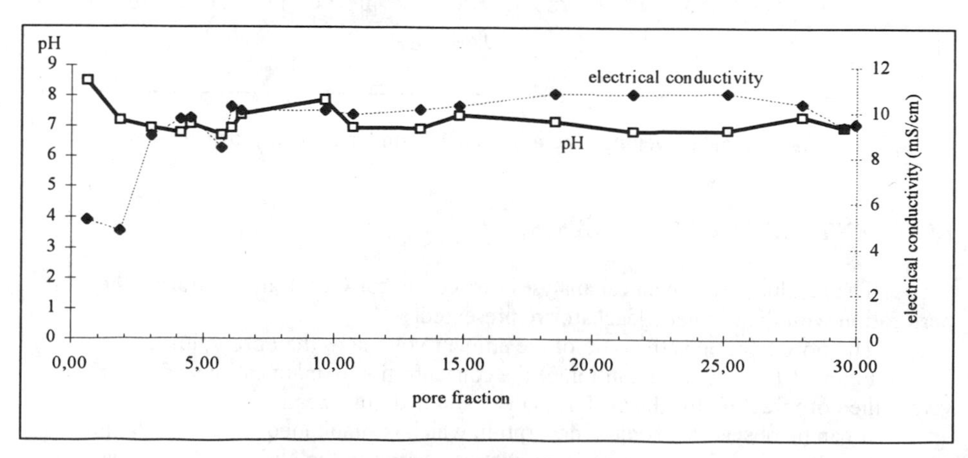

FIG. 11--Evolution of the pH and electrical conductivity versus pore fraction

Figure 12 shows the results of the effluent chemical analyses for GCL1a (partially hydrated with water).

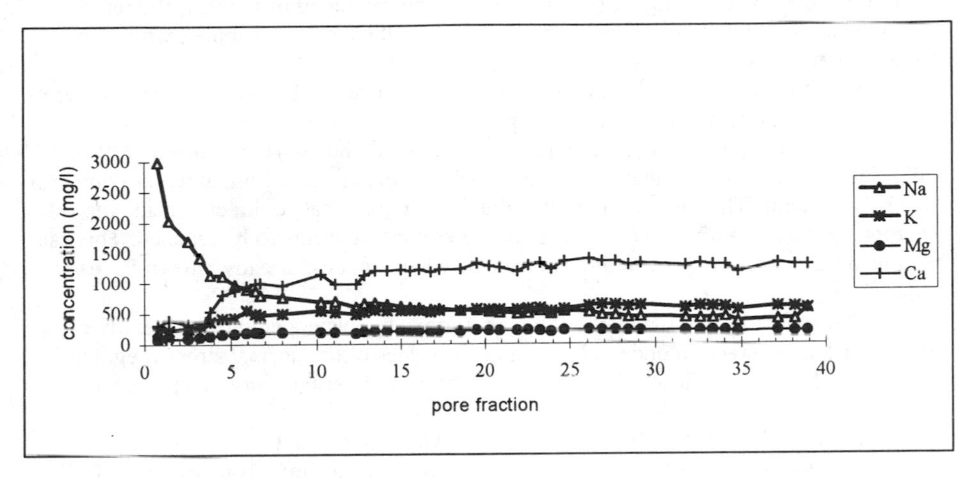

FIG. 12--Effluent cation concentration versus pore fraction for GCL1a.

The tendencies are the same for GCL1a: a desorption of the sodium and an adsorption of the calcium, potassium and magnesium. After a breakthrough of an effluent volume of 39 times the pore volume, the ionic exchanges seem to be stable.

The Na^+ concentration becomes equal with the Ca^{2+} for a pore fraction equal to 5.

The few tests analyzed todate are not enough to allow us definitive conclusions on the GCL permeability changes related to leachate. But, it can be noticed a sodium bentonite transformation, firstly into a calcium one, and then into a polycationic bentonite. The exchanges between Na^+ and Ca^{2+} take place quite rapidly after the contact with the leachate: 8 times the pore volume for the water saturated sample and 5 times for the water partially hydrated sample.

The global tendencies emphasized correspond with other research which is being carried out in the clay contamination field [2], [5].

CONCLUSIONS

The free swelling tests show an important decrease of the bentonite swelling in contact with the leachate. This seems to indicate that important permeability changes are to be expected when the GCL is saturated with leachate. It was confirmed by the tests on leachate saturated GCL, which permeability increased of three order of magnitude compared with the water permeability (5×10^{-8}m/s for the leachate saturated sample and 3×10^{-11}m/s for the water saturated one).

The tests show that the preliminary water saturation results in increase of only 2 times the leachate permeability compared with the demineralized water permeability.

The tests performed for a partially saturated, needle-punched GCL (GCL1) with a water content of 80%, show an increase of one order of magnitude of the leachate permeability compared to a water saturated GCL.

The effluent chemical analysis confirmed the transformation of the sodium bentonite into a polycationic one (Ca, K, Mg). At the beginning of the test, the removed sodium quantity is much more important for the partially hydrated sample then for the water saturated one.

The study of the initial hydration state revealed itself to be very important in terms of the real conditions of placement and exploitation.

The GCL's partially hydration methodology could be improved. Inspite of the difficulties, it allowed us to make a first evaluation of the impact of initial water content on the GCL behavior. The sample's partial hydration could be realize directly inside the cell by stopping the infiltration when the desired water content seems to be reached. This can be controlled by following the infiltration and swelling kinetics, already known for each material.

The apparatus and the methodology described by this paper allow the study of GCL - leachate interaction under a large range of values of the normal stress (beginning with a few kN/m^2), to follow the swelling, infiltration and permeability evolution with time.

The results were obtained by bringing the GCLs in contact with a smooth porous surface (geodrains, porous plate), but the apparatus will also allow us to study the GCL's behavior in contact with various granular materials.

ACKNOWLEDGMENTS

The authors acknowledge the ADEME (Agence de l'Environnement et de la Maîtrise de l'Energie, FRANCE) for this project initiative and for the partial support of the work.

We would like to thank Thierry WINIARSKI of the Environmental Sciences Laboratory - ENTPE (Ecole Nationale des Travaux Publiques de l'Etat, FRANCE) for performing the chemical analysis.

We also thank the GCL manufacturers (in alphabetical order): Colloid Environmental Technology Company (CETCO), Naue Fasertechnik and Volclay Limited, for supplying samples for this work.

REFERENCES

[1] Dakshanamurthy, V., "A new method to predict swelling using a hyperbolic equation", Geotechnical Eng., June 1978, vol.9, no.1, p. 29-38.

[2] Winiarski, T., "Analyse Sistémique de Fonctionnement de l'Interface Système Naturel et Système Antropisé:- Exemple de l'Interface Décharge/Aquifère", PhD thesis, Université Claude Bernard, Lyon, France, 1994.

[3] Madsen, F., Mitchell, J., "Chemical Effects on Clay Hydraulic Conductivity and Their Determination", Mitteilungen des Institutes für Grundbau und Bodenmechanik Eidgenossische Technische Hochschule Zürich, No. 135, Zürich, 1989.

[4] Egloffstein, T., Geosynthetic Clay Liners - Proceedings of an International Symposium Nürnberg, Germany, 14-15 April 1994, Balkema, Rotterdam, 1995.

[5] Cavalcante Rocha, Janaide, " Traitement des sables à la bentonite pour la constitution de la barrière de securité passive des centres de stockage de déchets", PhD thesis, INSA Lyon, France, 1995.

David E. Daniel[1], Stephen J. Trautwein[2], and Pernendu K. Goswami[3]

MEASUREMENT OF HYDRAULIC PROPERTIES OF GEOSYNTHETIC CLAY LINERS USING A FLOW BOX

REFERENCE: Daniel, D. E., Trautwein, S. J., and Goswami, P. K., **"Measurement of Hydraulic Properties of Geosynthetic Clay Liners Using a Flow Box,"** *Testing and Acceptance Criteria for Geosynthetic Clay Liners, ASTM STP 1308,* Larry W. Well, Ed., American Society for Testing and Materials, 1997.

ABSTRACT: An important property of geosynthetic clay liners (GCLs) is hydraulic conductivity. The authors describe and evaluate a new device, a GCL flowbox, that offers several advantages over large tanks or flexible-wall permeameters. The GCL flowbox can be used to test large-scale intact specimens, overlapped specimens with a full-width seam, and specimens subjected to environmental stresses such as freeze-thaw. A transparent, acrylic GCL flowbox was developed for testing at low normal stresses, and a metal version was developed for applying higher normal stresses. Results from the GCL flowbox compared favorably to results obtained with tanks and flexible-wall permeameters. It is concluded that the GCL flowbox provides a convenient and reliable testing methodology for verifying the hydraulic conductivity of both intact specimens and overlapped GCL panels.

KEYWORDS: hydraulic conductivity, permeability, bentonite, geosynthetic clay liner, GCL, overlap, freeze, thaw

---

The primary purpose of a geosynthetic clay liner (GCL) is to block the movement of fluids such as water, leachate, and gas. To accomplish this objective, GCLs must have low hydraulic conductivity. Fluids can move through GCLs in two ways (Fig. 1): (A) through the parent material, and (B) through overlaps of GCL panels. In a separate paper prepared for these proceedings, Daniel, Bowders, and Gilbert [1] address testing issues related to the hydraulic properties of the parent material. The subject of this paper is determination of the hydraulic characteristics of GCLs with a new device called a "GCL flowbox." The GCL flowbox is designed for testing intact GCL specimens as well as overlapped GCL panels.

---

[1] Professor, Department of Civil Engineering, University of Texas, Austin, TX 78712

[2] President, Trautwein Soil Testing Equipment Company, P. O. Box 31429, Houston, TX 77231

[3] Graduate Research Assistant, Department of Civil Engineering, University of Texas, Austin, TX 78712

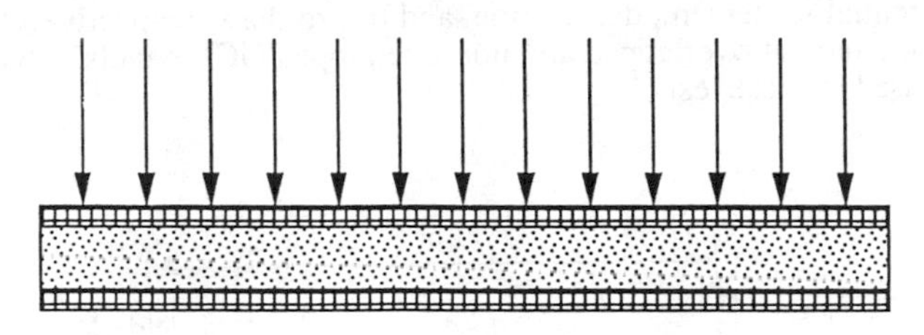

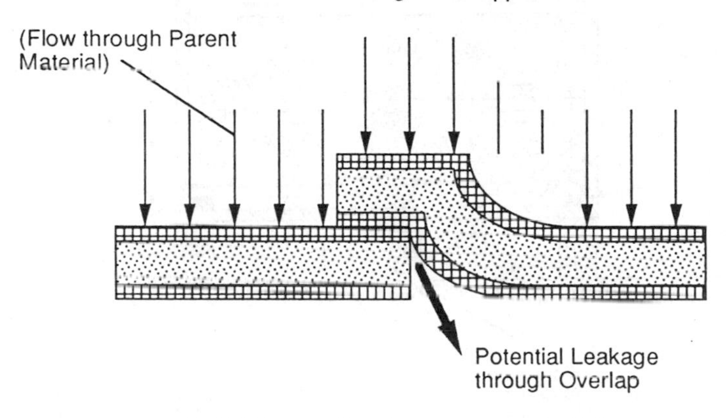

FIGURE 1 -- Flow Patterns through GCLs.

METHODS USED PREVIOUSLY FOR HYDRAULIC TESTING OF GCLs AND OVERLAPS

Hydraulic conductivity tests were performed on GCLs as early as the mid 1980s using flexible-wall permeameters [2]. Flexible-wall permeameters are by far the most commonly used device for measuring the hydraulic conductivity of parent GCL materials [1]. Flexible-wall permeameters are readily available and are convenient. Their principal limitations are that they are restricted to relatively small test specimens (diameters less than 100 to 300 mm) such that edge effects influence almost all testing, and they are difficult to use to study environmental effects (e.g., wet/dry cycles).

The first published attempt to determine the hydraulic performance of GCL overlaps was by Estornell and Daniel [3]. They used a steel tank that measured 2.4 m in length, 1.2 m in width, and 0.9 m in depth. A cross section through the tank is shown in Fig. 2. Tests were performed on overlapped and non-overlapped GCLs. For tests on overlapped panels, the overlap ran the full 2.4 m length of the tanks, and the GCLs were overlapped 37 mm (1.5 in.) to 150 mm (6 in.). The GCLs were covered with 300 to 600 mm of gravel, and were permeated by ponding 300 mm of water on the GCL. As shown in Fig. 2, outflow was collected from the base of the tank. Hydraulic conductivity was computed from the rate of outflow, using Darcy's law. A laboratory consolidometer was used to compress samples of GCL materials to a range of stress and to develop a plot of thickness versus compressive stress. The thickness of the GCL in the tank tests was estimated from the known compressive stress applied to the GCL in the tanks.

LaGatta [4], Boardman [5], and Hewitt [6] modified the tanks to evaluate the effects of differential settlement, desiccation, and freeze-thaw, respectively, on the hydraulic performance of overlapped and non-overlapped GCL panels. Overlaps up to 225-mm were used in these tests.

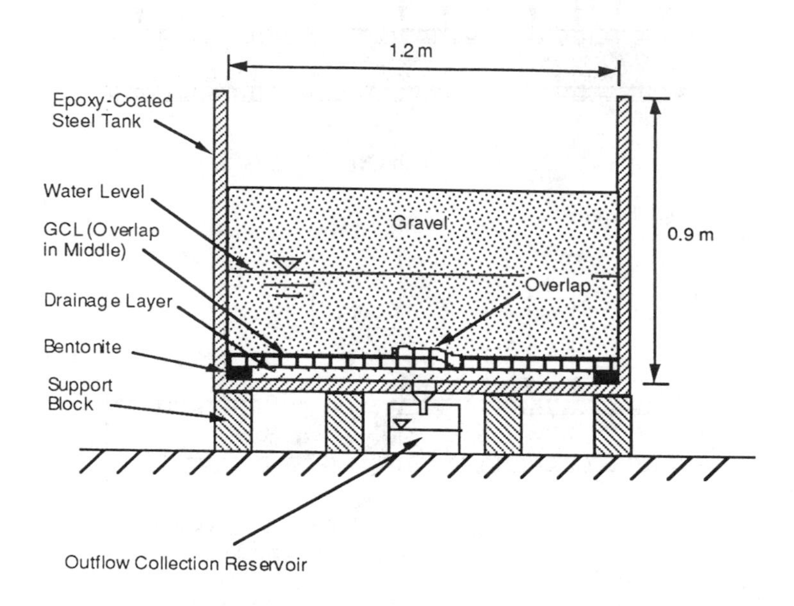

FIGURE 2 -- Tank used by Estornell and Daniel for testing overlaps [3].

The tanks used by Estornell, Boardman, LaGatta, and Hewitt have the following advantages: (1) full field-scale overlap widths are easily tested (field overlaps are usually 150 to 225 mm wide); (2) a significant length (2.4 m) of overlap is tested; and (3) the gravel overburden uniformly loads the overlap with known compressive stress. The principal disadvantages of the tanks are: (1) because the compressive stress cannot exceed about 15 kPa, the methodology is only applicable to testing at low compressive stress; and (2) the technique is not convenient because a lot of space is needed for the tanks and to store the gravel and because 2,700 kg (6,000 lb) of gravel must be placed in and removed from the tank for each test. The tanks are a useful research tool but are not practical for routine testing.

Cooley and Daniel [7] developed two techniques for evaluating the hydraulic integrity of overlapped GCL seams. One method involved a smaller tank than the one used by Estornell and Daniel, and the other involved a large flexible-wall permeameter.

The tank that Cooley and Daniel used is sketched in Fig. 3. The tank, constructed from PVC pipe and acrylic, was specifically developed for testing GCL overlaps and for determining the rate of flow through the overlap. Two pieces of GCL were cut and overlapped 75 mm in the center of the tank. Two acrylic strips, with bentonite on top to seal against the lower face of the GCL panels, were used to collect seepage that went directly through the overlap. Seepage through the non-overlapped portions of the GCL panels was collected through two separate outlet ports. The cell was successful in identifying and quantifying large flow rates through overlaps that were intentionally made to be faulty [7]. The principal limitations of this device are: (1) the circular cross section creates an overlap of variable width; and (2) the apparatus is limited to very low compressive stresses (< 15 kPa).

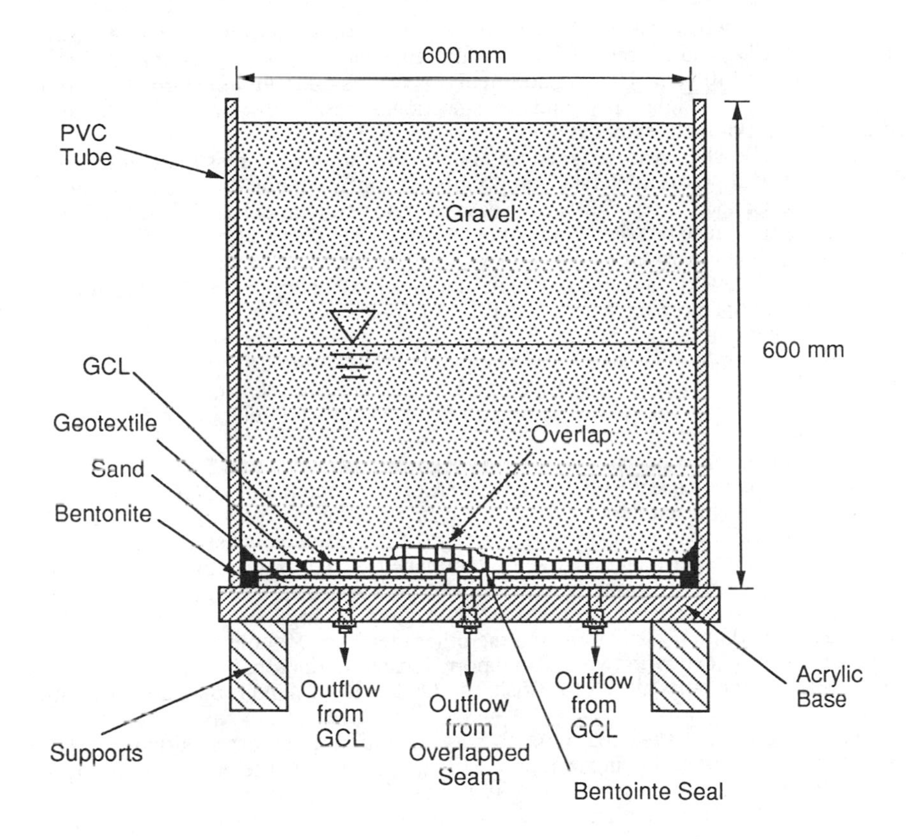

FIGURE 3 -- Tank used by Cooley and Daniel for testing overlaps [7].

Cooley and Daniel [7] modified the end pieces of a flexible-wall permeameter that can accommodate specimens with a diameter of 300 mm (12 in). The modified end pieces were designed to collect seepage in a manner similar to that shown in Fig. 3; there was a central collection zone to isolate flow that went through the overlap. The flexible-wall permeameter was used to permeate overlapped GCL panels at compressive stresses as high as 200 kPa. While this apparatus was successful in extending the range of compressive stress at which overlaps could be tested, the relatively small sample diameter restricted the width of overlaps to much less than actual overlap widths in the field.

In summary, the available techniques for testing GCLs and their overlaps each have advantages and disadvantages, but none is capable of conveniently testing large intact samples and full-width overlaps over a broad range of compressive stress. It was desired to develop a more convenient and flexible apparatus that would overcome these limitations. The apparatus that was developed is called a "GCL flowbox."

GCL FLOWBOX

The GCL flowbox is a box with a removable lid for measuring the rate of flow of water through GCLs, and the rate of flow through full-width overlaps, at compressive stresses as large as 140 kPa. The box nominally measures 600 mm in length, 350 mm in width, and 100 mm in height. Top- and cross-sectional views are shown in Fig. 4, and a schematic diagram of the assembled device is shown in Fig. 5.

The GCL flowbox was designed to: (1) provide larger test areas than typically used in flexible-wall permeameters; (2) apply higher confining pressure than used in gravel-filled tanks; and (3) test full-width seam overlaps conveniently. The size of the flowbox is a compromise between being able to test a large area and being able to conveniently apply an overburden pressure.

Two GCL flowboxes were constructed: a lower-cost acrylic version used for low normal stresses, and a more expensive metallic version (twice the cost of the acrylic version) for higher confining stresses. Both versions have the same dimensions and were constructed of 25-mm-thick plate stock. The parts of the acrylic box were glued together while those of the metal box were bolted together. The outer dimensions of the flowbox are 660 mm x 406 mm x 152 mm. The overall test area is rectangular and measures 610 mm x 356 mm.

The drainage zone beneath the GCL is segmented (Figs. 4 and 5) so that outflow from different areas beneath the GCL can be measured independently. The drainage zones are separated by a stainless steel or plastic strip that protrudes 3 mm above the base. A 25-mm-wide zone around the inside edge of the flowbox is used for monitoring the performance of the perimeter edge seal. Drainage holes from the edge seal area are used to identify any seepage from the periphery of the specimen, through the edge seal.

In addition to the edge zone, there are four other rectangular drainage zones (Fig. 4); three measure 305 mm x 152 mm, and the other measures 305 mm x 76 mm. Each drainage zone is provided with two outflow ports located in diagonally opposite corners. A groove between the ports allows for flushing. Outlet fittings and tubing connect with zero dead space.

The depth of the box is 102 mm. The configuration of a cross section for a typical test is constructed in the following sequence. First, a thin geotextile or similar drainage material is placed in the drain segments, between the 3-mm-high strips that separate the segments. Next, the test specimen(s) with outer dimensions equal to or slightly greater than the inside dimensions of the box is inserted. A bead of bentonite or other sealant is pushed into the GCL/flowbox interface to improve the seal. A 25-mm-thick layer of coarse material (sand or gravel) is then placed on top of the GCL. A geotextile is placed on the layer of coarse material. A molded rubber diaphragm is inserted. The geotextile beneath it serves to protect the diaphragm from puncture. The diaphragm flange is clamped between the box sides and lid (Fig. 5).

Confining pressure is applied to the GCL by first filling the space above the diaphragm with air or water. For low confining pressures, 0 - 15 kPa, pressure can be applied by attaching a standpipe to the pressure port and filling it with liquid as shown in Fig. 5. Alternatively, if a low air pressure is to be used, a pressure regulator may be employed. A water-filled manometer is used to measure the air pressure. The manometer also serves as a excess-pressure safety device — if too much air pressure is applied, water is blown out of the manometer. Higher pressure can be applied using a compressed gas source that interfaces with water, such that (for safety reasons) water fills the area above the flexible diaphragm in the GCL flowbox. The metal box has a limiting confining pressure of 200 kPa. The acrylic box has a limiting pressure of 15 kPa.

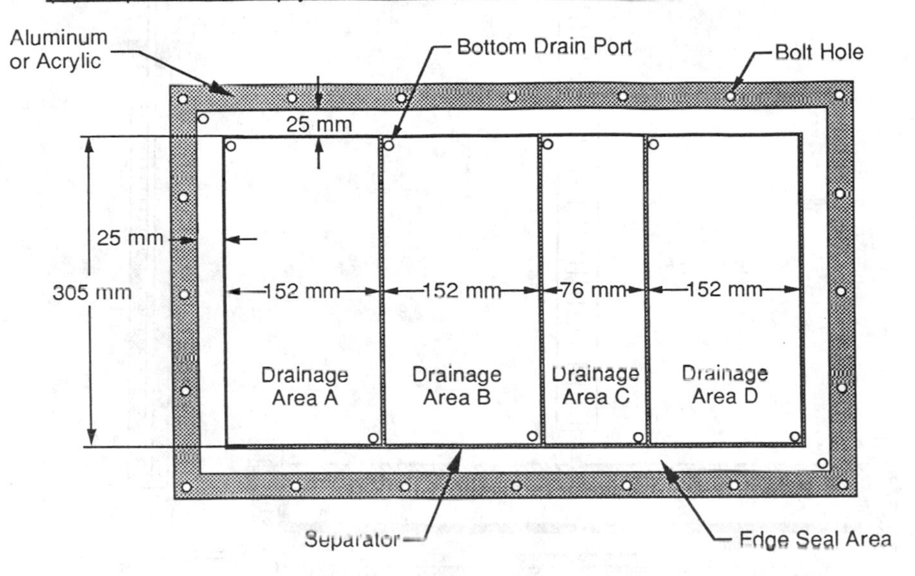

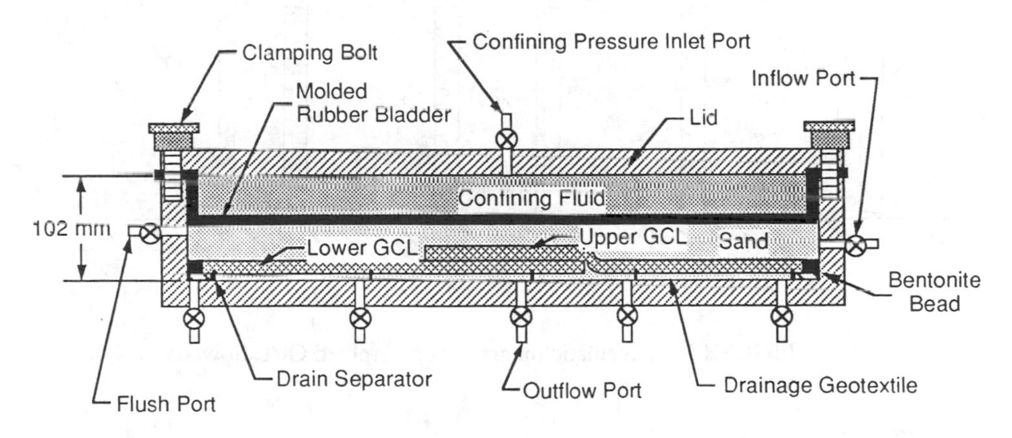

FIGURE 4 -- Top and cross-sectional views of GCL flowbox.

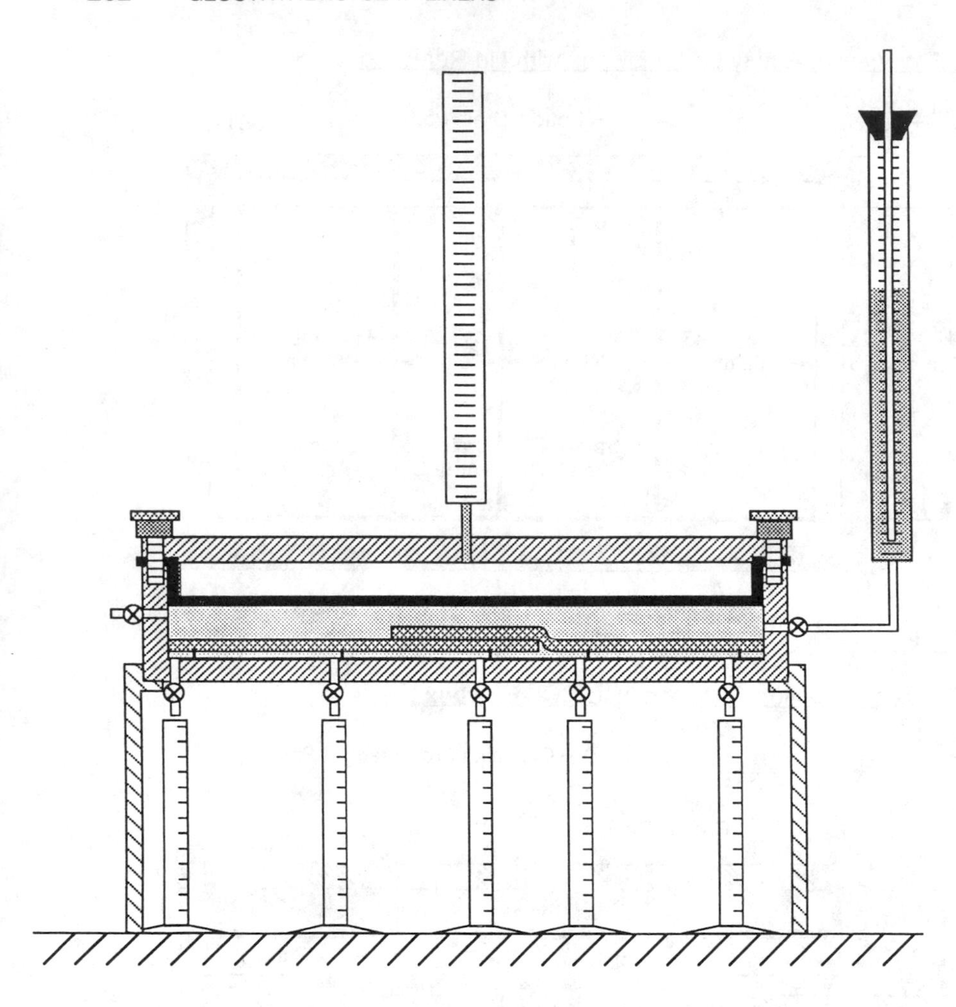

FIGURE 5 -- Schematic diagram of assembled GCL flowbox.

Permeant liquid is introduced through a side port. A second flush port on an opposite side facilitates saturation of the coarse material that overlies the GCL.

Permeation is accomplished by measuring the inflow and outflow quantities ove a known interval of time. Inflow can be controlled with constant head or falling head devices. Outflow is captured in graduated cylinders (Fig. 5). Swelling of the GCL can be monitored via a rod inserted through the confining pressure port. The lower end of t rod is attached to a small disc the rests on the top side of the diaphragm. However, the rod is not really necessary if the test will last for a sufficiently long period to ensure full hydration. The thickness of the test specimen is determined when the specimen is dismantled.

If a single GCL is used, outflow from section A, B, and D (Fig. 4A) can be compared to determine how uniform the flow is through the specimen. Flow from Section C can be compared to the other sections to see if there are variations in flux through the GCL. However, experience has shown that the flow rates through the various sections tend to be variable in short-term tests, and using the total flux through all four sections (A-D) provides the most meaningful indication of the hydraulic conductivity of the GCL. If two GCLs are used in an overlap configuration, Sections A and D can be compared for uniformity, Section B can be analyzed to determine if flow is reduced by adding a second layer of a GCL, and section C can be analyzed to determine if there is higher flow through between the interface of the GCLs or from the section of the upper GCL that bends over the edge of the lower GCL.

TYPICAL RESULTS

The plastic and metallic flowboxes have been used to measure the hydraulic conductivity of intact and overlapping GCLs. The tests were performed on a geotextile-encased, stitch-bonded GCL and a geotextile-encased, needlepunched GCL. The GCL test specimens were cut from randomly selected locations within large samples of the materials. The GCLs were set up in the flowbox either with no overlap (for control testing purposes) or with a 150-mm-wide overlap, which is the minimum overlap recommended by the manufacturer. Tests were performed in the clear acrylic GCL flowbox using the needlepunched GCL and in the metallic flowbox using the stitch-bonded GCL.

The GCLs were covered with a fine gravel, and the gravel was covered with a geotextile. Then a latex diaphragm was placed over the geotextile. Air pressure was used to apply confining pressures of 6 to 20 kPa. Permeation with de-aired tap water was initiated by imposing a head of approximately 200 mm of water on the top of the GCL specimens. Outflow was collected for a period of several weeks to several months. Hydraulic conductivity was computed from the total flux through all the outlet ports, and the thickness of the GCL as determined from tests performed over a range of vertical stresses in a consolidometer and verified at tear-down. Darcy's law was assumed to be valid. The hydraulic head was held nearly constant, and the tests were analyzed as constant-head tests.

Intact Specimens

A non-overlapping specimen of a geotextile-encased, needlepunched GCL was permeated in the clear acrylic GCL flowbox at a confining stress of 6 kPa. Results are summarized in Figure 6. The hydraulic conductivity remained essentially steady at 2×10^{-9} cm/s. This value is nearly identical to the values of hydraulic conductivity determined in small [7] and large [4-6] tank tests on this same material.

An intact specimen of a geotextile-encased, stitch-bonded GCL was permeated in the metallic GCL flowbox at an effective stress of 20 kPa. The variation in hydraulic conductivity over time is shown in Fig. 7. The test specimen was permeated for nearly two months, and the hydraulic conductivity was found to be about 5 to 6×10^{-10} cm/s, which is typical of values measured in the laboratory for this material at 20 kPa confining stress [3].

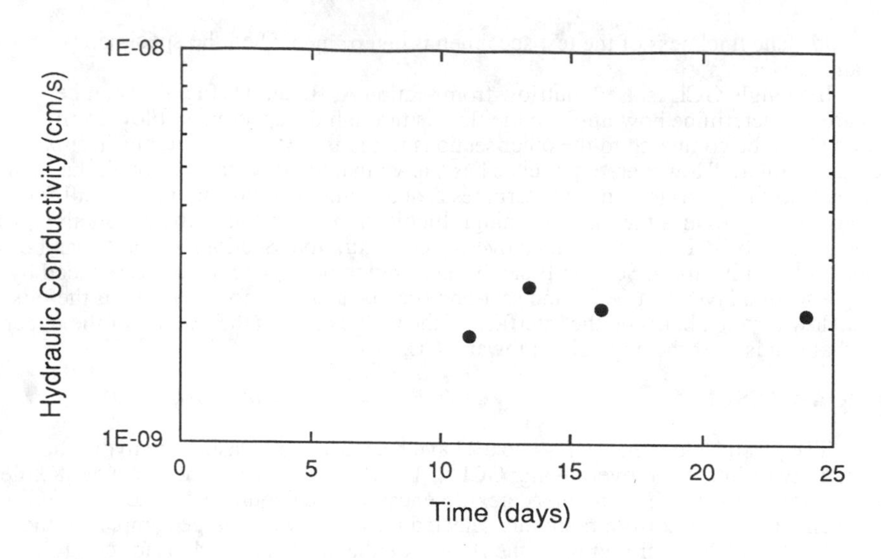

FIGURE 6 -- Results for intact needle-punched GCL tested in clear acrylic flowbox.

One of the ways that the GCL flowbox can be used is to study the effects of various factors on GCLs and their overlaps. To evaluate the usefulness of the GCL flowbox for this purpose, after permeation was complete, the GCL flowbox containing the test specimen was placed in a freezer and subjected to 10 freeze-thaw cycles. The test specimen was subjected to 2 freeze-thaw cycles per week, and the temperature inside the freezer was typically -20° C. Air pressure was continuously applied to the latex bladder to maintain an effective confining stress of 20 kPa. The GCL was not given access to extra water, but water remained in the gravel that overlay the GCL so the GCL could not desiccate. After the specimen had been subjected to 10 freeze-thaw cycles, the GCL flowbox was removed from the freezer, and the specimen was permeated. The results are shown in Fig. 8. The hydraulic conductivity after 10 freeze-thaw cycles was 1 to 2 x 10^{-9} cm/s, which is slightly higher than the value of 5 to 6 x 10^{-10} cm/s that was measured before freeze thaw. Hewitt [6] reported no significant change in the hydraulic conductivity of the same GCL.

Overlapped Specimens

Two tests have been performed on an overlapping, geotextile-encased, needle-punched GCL. Both tests were performed using the acrylic flowbox. In the first test, panels with the minimum recommended overlap width (150 mm) were set up, but no additional dry powdered bentonite was placed in the overlap. The manufacturer of this GCL recommends placing dry powdered bentonite in the overlap. The bentonite was intentionally left out to determine if the flowbox would correctly identify that the overlap was faulty. An effective confining stress of 6.2 kPa was applied, and then the GCL was permeated using a head of approximately 500 mm of water. Water flowed rapidly out of the drainage section located beneath the overlap. The rate of flow was 1.8 mL/s. The expected rate of flow, assuming that the overlap self-sealed, was 0.0001 mL/s. The overlap clearly did not seal properly, and the GCL flowbox successfully identified the faulty seal.

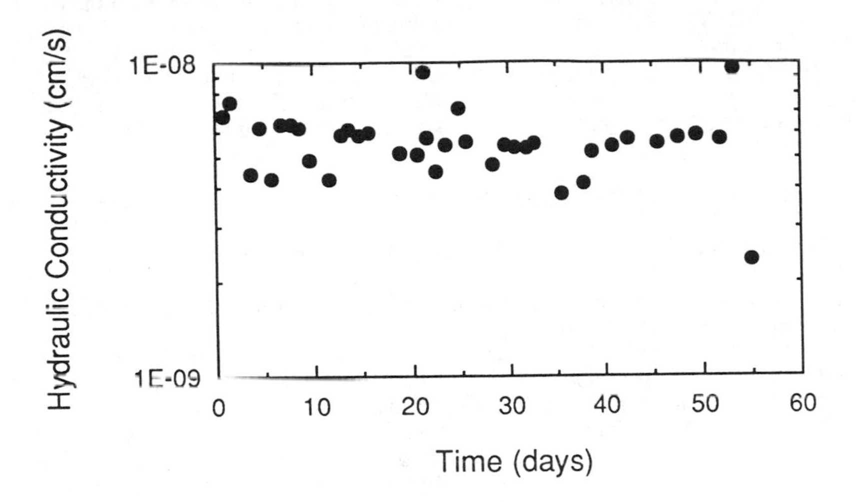

FIGURE 7 -- Results for intact stitch-bonded GCL tested in metallic GCL flowbox.

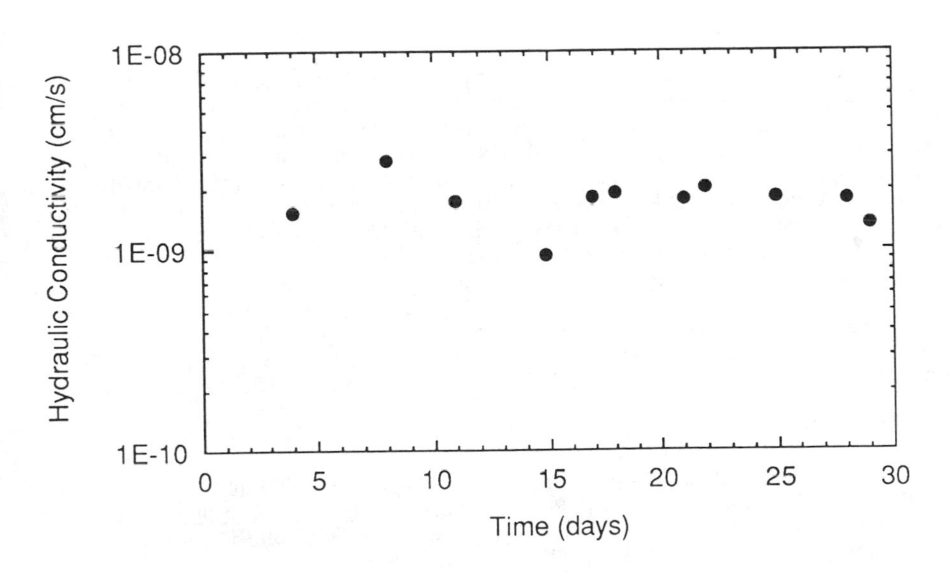

FIGURE 8 -- Results for intact stitch-bonded GCL tested in metallic GCL flowbox after 10 freeze-thaw cycles.

A second test was performed using 370 g/m of additional bentonite in the overlap. Once again, outflow only occurred from the drainage section located directly beneath the overlap. The outflow rate of 0.5 mL/s was much lower than in the first test, but still far larger than expected. Cooley and Daniel (1995) also had difficulty getting good self-seaming for this particular type of GCL in small tank tests. The cause for the incomplete self-sealing of the overlap is not known but may be related to the low confining pressure used in the tests.

POSSIBLE USES OF THE GCL FLOWBOX

The GCL flowbox might be used in several ways, including:

- For quality control testing at the manufacturing plant, to provide assurance that overlapped seams work properly;
- For conformance and quality assurance testing, to provide further assurance that GCL materials delivered to the job site will properly seal at overlaps;
- To study alternative sealing techniques for overlaps;
- To evaluate the self-sealing capabilities of prototype and new GCL products; and
- As a research tool, to evaluate the effects of parameters such as freeze-thaw, wet-dry, variations in temperature, effects of penetration by plant roots, composite action with a geomembrane, and others.

This paper was not intended to demonstrate all the ways in which the GCL flowbox might be used, but, rather, to document the design of the flowbox and to show typical results for a typical series of tests.

CONCLUSIONS

One of the important properties of geosynthetic clay liners (GCLs) is the hydraulic conductivity and self-seaming of overlaps. A GCL flowbox was developed that enables large samples of intact GCL materials or full-width seams to be tested over a broad range of confining stresses in a box that is relatively simple to set up and to use. A transparent box for use at low normal stresses, and a metallic box for higher normal stresses, were tested and found to provide results that compared favorably with results obtained from other, more cumbersome testing devices. One of the advantages of the GCL flowbox is that it permits convenient testing of GCLs and their overlaps over a range of environmental conditions, e.g., freeze-thaw. Although the GCL flowbox is new and has only been used in a limited way, it is far simpler and more convenient than any other testing device that has been employed for permeating large GCL samples or for testing GCL overlaps. It is concluded that the GCL flowbox provides a simple and effective means for testing large samples of GCLs and for testing GCL overlaps.

REFERENCES

1. Daniel, D.E., Bowders, J.J., and Gilbert, R.B., "Laboratory Hydraulic Conductivity Testing of GCLs in Flexible Wall Permeameters," Testing and Acceptance Criteria for Geosynthetic Clay Liners, ASTM STP 1308, Larry W. Well, Ed., American Society for Testing and Materials, Philadelphia, 1996.

3. Estornell, P.M., and Daniel, D.E., 1992, "Hydraulic Conductivity of Three Geosynthetic Clay Liners," Journal of Geotechnical Engineering, 118(10): 1592-1606.

4. LaGatta, M.D., 1992, "Hydraulic Conductivity Tests on Geosynthetic Clay Liners Subjected to Differential Settlement," Master of Science Thesis, University of Texas, Austin, Texas, 120 p.

5. Boardman, B. T., 1993, "The Potential Use of Geosynthetic Clay Liners as Final Covers in Arid Regions," Master of Science Thesis, University of Texas, Austin, Texas, 110 p.

6. Hewitt, R.D., 1994, "Hydraulic Conductivity of Geosynthetic Clay Liners Subjected to Freeze/Thaw," Master of Science Thesis, University of Texas, Austin, Texas, 103 p.

7. Cooley, B.H., and Daniel, D.E., 1995, "Seam Performance of Overlapped Geosynthetic Clay Liners," *Geosynthetics '95*, Industrial Fabrics Association International, St. Paul, MN, pp. 691-705.

David E. Daniel[1], John J. Bowders[2], and Robert B. Gilbert[3]

LABORATORY HYDRAULIC CONDUCTIVITY TESTING OF GCLs IN FLEXIBLE-WALL PERMEAMETERS

REFERENCE: Daniel, D. E., Bowders, J. J., and Gilbert, R. B., **''Laboratory Hydraulic Conductivity Testing of GCLs in Flexible-Wall Permeameters,''** *Testing and Acceptance Criteria for Geosynthetic Clay Liners, ASTM STP 1308,* Larry W. Well, Ed., American Society for Testing and Materials, 1997.

ABSTRACT: One of the important properties of geosynthetic clay liners (GCLs) is hydraulic conductivity. In the laboratory, hydraulic conductivity of GCLs is typically measured in flexible-wall permeameters. The most important variables in hydraulic conductivity testing of GCLs are addressed, including (i) trimming the GCL specimen; (ii) determining the thickness of the specimen; (iii) selecting the effective stress; (iv) selecting the hydraulic gradient; and (v) selecting the first wetting liquid and permeant liquid. A round-robin testing program was conducted in which 18 laboratories independently measured the hydraulic conductivity of a GCL that was permeated with water. The test specimens all came from the same GCL sample. The coefficient of variation, accounting for all sources of variability, was 42%. For experienced laboratories, this value reduced to 35%, which was identical to the variation in quality control tests performed over a 7-month period by the manufacturer on a variety of GCL samples. The round-robin test results are encouraging; there was less variability than might be expected, considering the difficulty in accurately measuring the hydraulic conductivity of relatively impermeable materials such as GCLs.

KEYWORDS: hydraulic conductivity, permeability, bentonite, geosynthetic clay liner, GCL, flexible-wall permeameter, precision, bias, variability

---

Hydraulic conductivity is the coefficient of proportionality in Henry Darcy's 1856 law, which can be written as:

$$q = k\, i\, A = k \frac{\Delta H}{L} A \qquad (1)$$

---

[1] Professor, Department of Civil Engineering, University of Texas, Austin, TX 78712

[2] Research Scientist, University of Texas, Department of Civil Engineering, Austin, TX 78712

[3] Assistant Professor, Department of Civil Engineering, University of Texas, Austin, TX 78712

where q is the rate of flow (cm^3/s), k is the hydraulic conductivity (cm/s), i is the hydraulic gradient (dimensionless and equal to $\Delta H/L$, ΔH is the head loss across specimen (cm), L is the length of specimen (cm), and A is the cross-sectional area of specimen perpendicular to the direction of flow (cm^2).

The hydraulic conductivity in Darcy's law depends not only upon the properties of the porous medium but also upon the density and viscosity of the permeating liquid [1]. Darcy's law may be written in an alternative form using intrinsic permeability, which is a property only of the porous medium [2].

One of the most important characteristics of geosynthetic clay liners (GCLs) is hydraulic conductivity. The hydraulic conductivity to water, as well as the conductivity to chemicals or leachates, is normally measured in laboratory permeameters. The purpose of this paper is to describe the current state-of-the-practice for laboratory hydraulic conductivity testing of GCLs, and to present the results of a round-robin testing program. The scope of the paper is limited to testing relatively small-diameter, laboratory test specimens. Larger tests, e.g., on GCL overlaps, are normally conducted with larger scale equipment [3], which is beyond the scope of this paper.

HYDRAULIC CONDUCTIVITY CELLS

Numerous variations of hydraulic conductivity cells are available for permeating porous media. The cells may be divided into two broad categories: rigid-wall and flexible-wall cells [2]. Flexible-wall permeameters are almost universally used for permeating GCLs because flexible-wall cells permit accurate control over all the stresses acting on the specimen and, more importantly, allow the test specimen to be fully saturated using "back-pressure." Back-pressure refers to pressurizing both the influent and effluent ends of the test specimen; the pressure compresses and dissolves gas bubbles.

A schematic diagram of a flexible-wall permeameter is shown in Fig. 1. The upper and lower surfaces of the test specimen are confined with porous disks (although some non-woven, needle-punched geotextiles have also been used). A sheet of filter paper separates the specimen from the porous end pieces to prevent plugging of the porous elements from bentonite. The porous end pieces are confined on top with a top cap and on the bottom with a base pedestal. A latex membrane confines the sides. Two drainage lines to the top and bottom end pieces are recommended for flushing air out of the lines. The cell is filled with water and pressurized to press the latex membrane against the test specimen and thereby minimize or eliminate sidewall leakage. The cell may be filled with some other fluid besides water for special tests. For instance, if the test specimen is to be frozen, a mixture of ethylene glycol (antifreeze) and water may be used for the cell fluid.

When the permeating liquid is a chemical or waste fluid, three potential problems must be addressed: (1) compatibility of the liquid with the components of the flexible-wall permeameter; (2) compatibility of the liquid with the latex membrane; and (3) diffusion of the permeating fluids, or solutes in the fluid, through the confining membrane and into the cell liquid. The first problem is the easiest to solve; the cell is built with materials such as stainless steel and Teflon® that are compatible with the waste liquid. The second problem can be solved in two ways: (1) another flexible membrane material besides latex (e.g., neoprene) can be used, or (2) the test specimen can first be wrapped with a thin film of Teflon® (the authors use a wide roll of Teflon® tape with an adhesive on one side), and then a latex membrane can be placed over the Teflon®. The authors have had good success with the Teflon® wrapping. The problem of diffusion through the membrane is solved by placing a diffusion barrier (Teflon® film or aluminum foil, for instance) between the membrane and test specimen or between two membranes that encase the specimen.

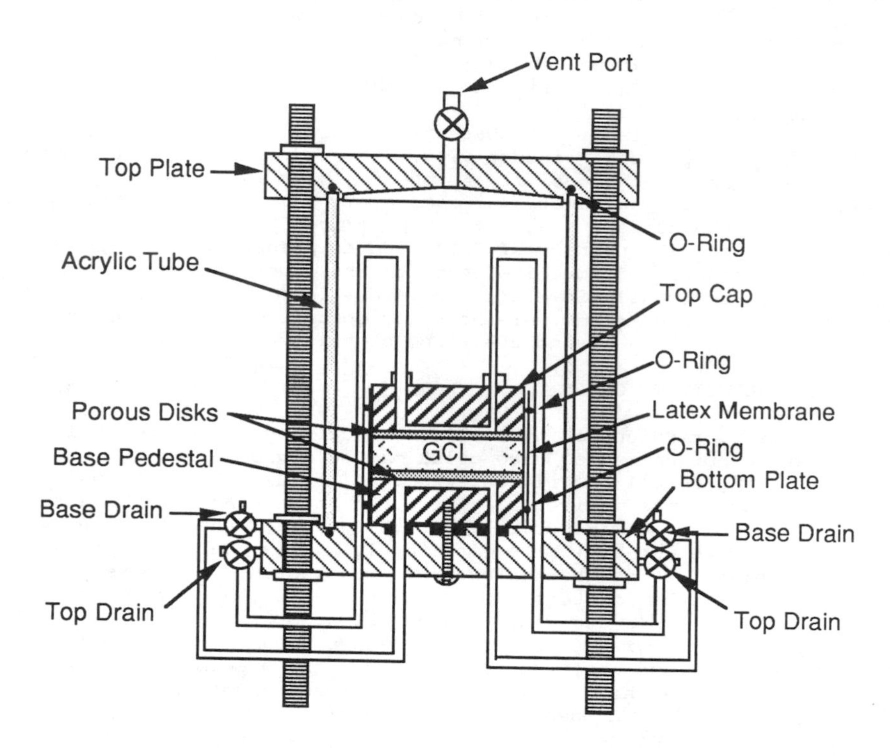

FIGURE 1 -- Flexible-wall permeameter.

Additional information on flexible-wall cells and testing with flexible-wall permeameters is given by [2, 4-7] and by, "Measurement of Hydraulic Conductivity of Saturated Porous Materials Using a Flexible Wall Permeameter," ASTM D5084. ASTM Committee D35 has recently passed a standard for measuring the rate of water flow through GCLs, "Standard Test Method for Measurement of Index Flux through Saturated Geosynthetic Clay Liner Specimens using a Flexible Wall Permeameter," ASTM D5887.

Size of Test Specimen

The test specimen should be as large as practical [2]. The authors have frequently tested GCL specimens that measured 100 to 300 mm in diameter. They and others (e.g., [2]) have consistently found no relationship between hydraulic conductivity and the diameter of the test specimen for specimens larger than 100 mm in diameter. Thus, 100-mm diameter is sufficiently large to obtain representative results. Even smaller specimens, e.g., 75 mm diameter, probably yield the same hydraulic conductivity as 100-mm-diameter specimens, but comparative data are not available.

B-Coefficient

With a flexible-wall permeameter, saturation of the test specimen can be verified prior to permeation by measuring the B coefficient [8]. The B coefficient is defined as follows:

$$B = \frac{\Delta u}{\Delta \sigma} \tag{2}$$

where Δu is the change in pore water pressure in the test specimen (kPa), and $\Delta \sigma$ is the change in all-around confining pressure (kPa). If the GCL is completely saturated and no air is present, the pore water pressure will change by the same amount as the cell pressure and B will equal to 1.0. The greater the amount of air present in the test specimen (or porous disks or drainage tubes), the lower will be the B coefficient. A final degree of saturation ≥ 95% is specified in ASTM D5084, and most laboratories strive for a B coefficient ≥ 0.95.

HYDRAULIC CONTROL SYSTEMS

Hydraulic control systems are needed to control the delivery of fluid to a permeameter. In order to determine the hydraulic conductivity from Eq. 1, one must measure the flow rate (q) and head loss (ΔH). Several types of test are possible:

- Constant-head test in which the head loss is kept constant and the corresponding rate of flow is measured.
- Varying-head test in which the head loss declines with time in a measured manner and the rate of flow is computed from the change in water level and the area of the tube in which the head falls.
- Constant-rate-of-flow test in which the rate of flow is kept constant and the corresponding head loss is measured.

The main advantage of a constant head test is the simplicity of calculation of hydraulic conductivity. However, the various designs that are available for constant-head hydraulic systems [2] are not particularly convenient for testing under backpressure and are not routinely employed in commercial testing equipment. Virtually all commercial hydraulic conductivity tests on GCLs are falling-head tests because the apparatus for falling-head tests is simpler. However, as indicated in ASTM D5084, if the head does not change very much (less than 10%), then the test may be treated as a constant-head test without introducing significant error.

A variable-head test is one in which the water level in the influent or effluent reservoir (or both) changes during permeation. Two types of variable-head tests are possible: (1) a falling-headwater, constant-tailwater-pressure test, and (2) a falling-headwater, rising-tailwater-pressure test. The test with the rising-tailwater-pressure test tends to be more convenient for testing under backpressure and, therefore, for testing GCLs. As discussed by [9], the equations that are used for computing hydraulic conductivity are different for the conditions of a constant tailwater pressure and a rising tailwater pressure. If the influent and effluent standpipes in a rising-tailwater-pressure test have the same area, then the equations differ by a factor of 2.

The constant-rate-of-flow test is performed by pumping permeant liquid through the test specimen at a controlled rate and measuring the pressure drop across the test specimen. In the simplest arrangement, a differential electronic pressure transducer is connected to a strip-chart recorder. Once the flow rate and pressure drop become steady, the test is complete (unless a chemical or waste liquid is being used, in which case permeation continues until other termination criteria are met, as discussed later). The advantages of the constant-rate-of-flow test are: (1) equilibration can be achieved very rapidly if the test specimen is saturated with water, which minimizes testing time; (2) the entire test, including control of the test, is easily automated; and (3) when the permeant liquid is a chemical or waste liquid, the flow pump can be set to deliver a known amount of liquid in a given amount of time, e.g., 2 pore volumes of liquid in 4 weeks, which can be helpful in making sure that sufficient throughput of liquid occurs within the time constraints of a project. Disadvantages of the constant-rate-of-flow test are high equipment cost and the possibility of developing extremely large hydraulic gradients if too large a flow rate is used.

An alternative hydraulic control system is shown in Fig. 2. With this system, permeation is effected with a falling headwater and rising tailwater, but the system is designed to maintain a constant volume in the test specimen. The closed manometer ensures that inflow and outflow are equal, which in turn ensures that a saturated test specimen does not change volume. If the soil does not change volume when permeation is initiated, equilibrium is reached sooner and testing time is minimized. This type of hydraulic control system is particularly advantageous for testing extremely low-hydraulic-conductivity materials, such as GCLs.

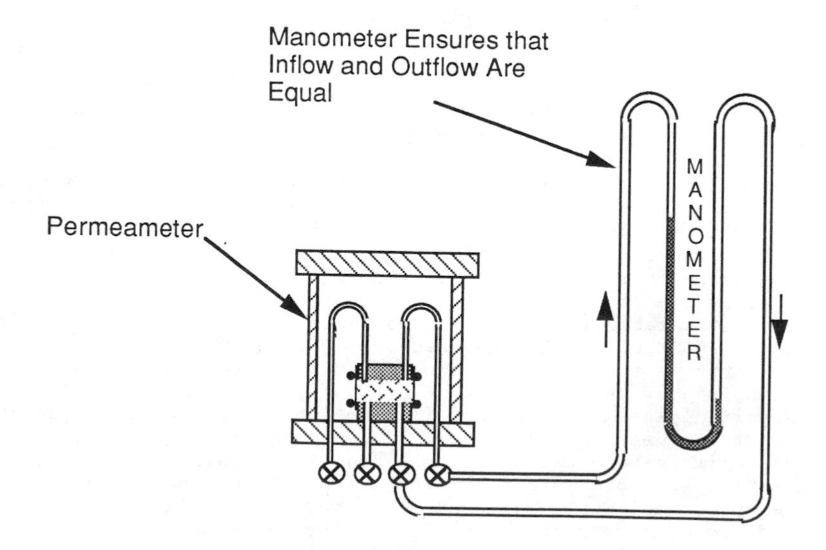

FIGURE 2 -- Hydraulic system with closed manometer to maintain a constant volume in the test specimen.

Interface between Pressure Panels and Permeameter

Most commercial hydraulic systems involve use of pneumatic pressure panels. Reservoirs are filled with water and are pressurized with air. When a hydraulic conductivity test is to be performed with a chemical or waste liquid, it is convenient to provide an interface between the permeant liquid and the water in the reservoir to avoid contaminating the panel board apparatus.

The interface that is commonly used is a bladder accumulator (Fig. 3). The accumulator consists of a flexible diaphragm ("bladder") contained within a chamber. Water is contained on one side of the bladder and the permeant liquid on the other. Care must be taken to make sure that the bladder and all materials on the permeameter side of the bladder are chemically compatible with the permeant liquid. Stainless steel is the principal material used on the permeameter side of the bladder. Acrylic is typically used on the panel side, since this side is filled only with water. The bladder must be chemically compatible with the permeant liquid. Butyl rubber is the most commonly used material for the bladder. Also, removing all the air from the accumulator can be difficult and time consuming.

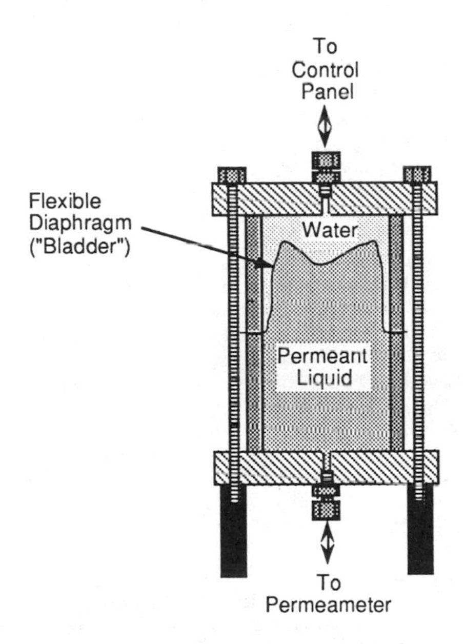

FIGURE 3 -- Bladder accumulator.

PERMEANT LIQUID

Hydraulic conductivity tests are usually performed either with water or with a chemical or waste liquid as the permeant liquid. Chemicals require special considerations in terms of equipment design.

Water

The most important characteristics of the permeant water are the amount of dissolved air in the water, the type and concentration of electrolytes, turbidity, nutrient content, and population of microorganisms. It is best to permeate soils with deaired water. "Deaired water" means water from which dissolved air has been substantially removed. A dissolved oxygen meter may be used to confirm proper deairing. At atmospheric pressure, water will contain approximately 8 mg/L of dissolved oxygen. Properly deaired water will contain less than 1 to 2 mg/L of dissolved oxygen. Water is deaired by boiling the water or by placing the water in a container that is subjected to vacuum. With a vacuum chamber, it is best to spray the water into the evacuated chamber in a fine mist (e.g., with a spray paint nozzle) to expose a large surface area of the water to vacuum and to facilitate rapid deairing. If the permeant water is boiled, care must be taken not to increase the salinity too much.

Three types of deaired water are typically used in laboratory hydraulic conductivity tests on GCLs:

1. Tap water, as recommended by ASTM D5084.
2. Distilled water, as required by Geosynthetic Research Institute (GRI) Method GCL-2.
3. Deionized water, as required by ASTM D5887.

Tap water was specified in ASTM D5084 in an attempt to match the chemistry of the permeating liquid as closely as possible to real-world soil water. It was recognized by the writers of ASTM D5084 that tap water would never precisely simulate soil water in the field, but tap water contains some ions, and it was believed that tap water would be a better choice than distilled water, which is free of any ions. Perhaps the most negative aspect of recommending tap water is the fact that the chemistry of tap water varies considerably from one location to another.

GRI Method GCL-2 requires distilled water. Distilled water yields a lower hydraulic conductivity for clays than tap water [1], but tap water varies from one location to another. But GRI Method GCL-2 was intended as a conformance test (not a performance test, which was the intent of ASTM D5084), and consistency of results between laboratories is critical in a conformance test. Thus, the use of distilled water over tap water for GRI Method GCL-2 is justified on the basis of minimizing scatter in test results between laboratories.

Chemicals and Waste Liquids

Permeation with chemicals or waste liquids presents a number of additional challenges, which include health and safety considerations, material compatibility concerns, cross-contamination potential from one test to the next, changes in chemistry of the influent liquid, and monitoring of the chemistry of the effluent liquid.

Health and safety considerations often require one to set up the entire hydraulic conductivity system in an environmentally controlled room with separate air conditioning, constant replenishment of air, and appropriate emergency and first-aid equipment on hand.

Material compatibility issues are usually addressed by using stainless steel or other chemically resistant materials for those components that will come into contact with the permeant liquid. Bladder accumulators (Fig. 3) help to reduce the number of pieces of equipment that will contact the permeant liquid. Cross-contamination must be dealt with by thoroughly cleaning the equipment after each use.

Typically, with GCL testing involving chemicals or leachates, the main issue is changes in hydraulic conductivity after several pore volumes of flow. Rarely is the chemistry of the effluent liquid monitored, but effluent monitoring may be necessary to verify full breakthrough of key chemicals or to determine the effective porosity of the GCL. In nearly all cases, hydraulic conductivity is measured and plotted versus pore volumes of flow and/or time of permeation.

PREPARATION OF TEST SPECIMENS

For geotextile-encased GCLs, the main challenge in preparing a test specimen is preventing loss of bentonite along the edges of the GCL. If steps are not taken to prevent the problem, dry bentonite powder tends to fall out of the edges of the material, as shown in Fig. 4. The geotextiles along the edge can pinch together, with no bentonite separating the geotextiles. This leads to measurement of erroneously high hydraulic conductivity. In the authors' laboratory, this happened several years ago during the first attempt to test this type of GCL, and the measured hydraulic conductivity was 1 x 10^{-6} cm/s. It was obvious that this value of hydraulic conductivity was much too high to be representative.

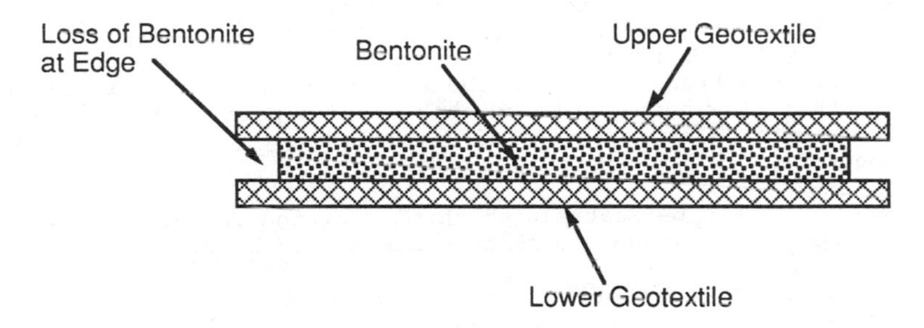

FIGURE 4 -- Problem with bentonite falling out from edge of GCL.

The problem with bentonite falling out from the edges of geotextile-encased GCLs is easily solved by scribing a mark on one geotextile indicating the circular cross section to be cut out, and then applying a small amount of water along the marked circle with a squirt bottle. This moistens the bentonite in the area to be cut, which prevents loss of bentonite. More water is added after the upper geotextile is cut, if necessary. This same type of problem can occur, and can be resolved in the manner described, with any type of GCL for which no adhesive is added to the bentonite. In instances when bentonite does fall out from the edges of the geotextiles, moistened bentonite should be packed between the geotextiles.

With a geomembrane-supported GCL, there is no point in permeating the GCL with the geomembrane backing in place because the geomembrane has a water vapor transmission rate that is too low to measure accurately with a flexible-wall permeameter. The bentonite component of the GCL can be tested by carefully cutting away the geomembrane with a knife as described in [10].

TERMINATION CRITERIA

"Termination criteria" refers to criteria for terminating a hydraulic conductivity test. Different criteria apply to tests with water and with chemicals or waste liquids.

When GCLs are permeated with water, the following criteria are recommended for determining when a test is complete:

- The rates of inflow and outflow should be reasonably equal. The bentonite in GCLs will continue to absorb water and hydrate for weeks, and setting too tight a requirement for equal inflow and outflow will guarantee very long testing times. Experience has shown that a good target value is for the ratio of outflow to inflow (q_{out}/q_{in}) to be between 0.75 and 1.25.

- The hydraulic conductivity should be reasonably steady. It is good practice to plot hydraulic conductivity versus time or pore volumes of flow. (One pore volume of flow has occurred when the cumulative quantity of inflow equals the computed volume of void space in the bentonite.)

- Enough data points should be collected to ensure that the test has lasted long enough to achieve representative results. Typically, it is recommended that at least 3 to 4 consecutive hydraulic conductivity points should yield computed values that do not scatter more than ± 25% from the average value. Our experience has been that flow needs to be measured over a period of least 8 hours (and sometimes at intervals of several days) in order to have enough flow to measure it accurately.

When GCLs are permeated with chemicals, the same requirements for water apply, but some additional controls should be established:

- Permeation should continue until at least 2 pore volumes of the permeant liquid has flowed through the GCL. This will ensure that the remnant soil water has been flushed out of the test specimen.

- Permeation should continue until the chemical composition of the effluent liquid is similar to that of the influent liquid and until the key compounds that might alter hydraulic conductivity appear in the effluent liquid in concentrations that are similar to the influent liquid [11, 12].

KEY TESTING VARIABLES

Effective Compressive Stress

The hydraulic conductivity of a GCL decreases with increasing effective compressive stress [2, 13]. For GCLs tested at the lowest possible compressive stress (5 to 10 kPa), the hydraulic conductivity is often in the mid 10^{-9} cm/s range, but for GCLs tested at compressive stresses on the order of 300 kPa, the conductivity is about 1.5 orders of magnitude lower, in the low 10^{-10} cm/s range. No specification of hydraulic conductivity of a GCL is meaningful without stating the effective stress at which the hydraulic conductivity is measured. ASTM D5084 does not specify the effective compressive stress (the individual requesting the test is expected to specify the appropriate value for the application of interest), but GRI Method GCL-2 does specify a maximum compressive stress of 69 kPa at the effluent end of the specimen. ASTM D5887 an index flux test, requires a cell pressure of 550 kPa, an effluent backpressure of 515 kPa, and in influent backpressure of 530 kPa, thereby establishing a maximum effective stress of 35 kPa.

Hydraulic Gradient

Application of an excessively large hydraulic gradient is undesirable [2]. In general one should try to use a hydraulic gradient as close as possible to the value expected in the field.

ASTM D5084 recommends that the hydraulic gradient for testing very low-hydraulic-conductivity materials not exceed 30. GRI Method GCL-2 recommends a 35 kPa pressure difference across the GCL, which creates a very large hydraulic gradient of (depending on the thickness of the GCL) about 400. GRI Method GCL-2 is a conformance test that is intended to be used for fairly rapid verification of low hydraulic conductivity under a specified, arbitrary set of conditions. Similarly, ASTM D5887 specifies a 15 kPa pressure difference across the specimen in order to quickly measure an index flux value. For performance testing, a much lower hydraulic gradient (like that recommended in ASTM D5084) is suggested.

First Wetting Liquid

For permeation of GCLs with chemicals or waste liquids, the key testing variables are compressive stress (discussed earlier) and the first wetting liquid. GCLs that are first hydrated with water and then permeated with chemicals or waste liquids tend to have lower hydraulic conductivity than GCLs permeated directly with the chemical or waste liquid [12-15]. For example, Ruhl [13] permeated a GCL with simulated municipal solid waste (MSW) leachate. When the GCL was permeated with water and then with the simulated MSW leachate, the hydraulic conductivity was 2×10^{-9} cm/s. When the same GCL material was permeated directly with the simulated MSW leachate, the hydraulic conductivity was 2×10^{-5} cm/s, which is 10,000 times larger than the value obtained when the first wetting liquid was water rather than leachate. Water causes bentonite to hydrate and swell, and many chemicals (especially those rich in multivalent cations, such as calcium) cause less swelling.

Direct permeation with the chemical or waste liquid is typically the most critical case. If it is possible that the GCL will be initially hydrated in the field with the chemical or waste liquid of concern, then the first wetting liquid in the laboratory should be the relevant chemical or waste liquid, rather than water.

Determining the Thickness of the Test Specimen

For geotextile-encased GCLs, an important testing consideration is determination of the thickness of the test specimen. The issue is whether to include the thickness of the geotextiles in the thickness of the test specimen. In principle, the geotextiles should not be included because the bentonite, not the geotextiles, is providing the resistance to fluid flow. However, it is very difficult to determine accurately the thickness of the bentonite, and even if it could be determined, bentonite tends to swell into the geotextiles, making the actual boundary of the bentonite diffuse and indeterminate. Thus, the only practical option is to measure the gross (total) thickness of the GCL specimen, including the thickness of the geotextiles. The only problem with this practice is that the actual thickness is less than assumed, and as a result, the actual hydraulic conductivity is less than calculated. Also, the thicker the geotextiles, the greater the calculated hydraulic conductivity. It is inappropriate to "penalize" a GCL (i.e., report to high a hydraulic conductivity) simply because it has thicker geotextiles than other GCLs. ASTM D5887, an index flux test, avoids this problem by not attempting to determine hydraulic conductivity, but just flux under a specified gradient, which can be determined without measuring the thickness of the test specimen.

ROUND-ROBIN TESTING PROGRAM

To the author's knowledge, no round-robin testing program involving hydraulic conductivity testing on clay or GCLs has ever been described in the literature. In the mid 1980s, the U.S. Environmental Protection Agency sponsored a round-robin testing program for flexible-wall hydraulic conductivity testing of compacted clay, but the results were never published. Very large variability in the test results was one reason cited for not releasing the results.

A round-robin testing program was initiated to evaluate the variability of industry-standard procedures for measuring the hydraulic conductivity of GCLs. Eighteen laboratories, listed in Table 1, participated in the round-robin program.

TABLE 1 -- Laboratories participating in round-robin program.

Laboratory	Location	Responsible Individual
Advanced Terra Testing	Lakewood, CO	Chris Wienecke
Ardaman & Associates	Orlando, FL	Shawkat Ali
Colorado State Univ.	Fort Collins, CO	Charles D. Shackelford
EMCON	Tuxcedo, NY	Rasheed Ahmed
French & Parrello Assc.	Holmdel, NJ	David Calnan
GeoSyntec Consultants	Atlanta, GA	Nadar Rad
GeoSystems Consultants	Fort Washington, PA	Craig Calabria
NTH Consultants, Ltd.	Farmington Hills, MI	Jenghwa Lyang
Rensselaer Poly. Inst.	Troy, NY	Thomas F. Zimmie
RUST Environ. & Infra.	Baton Rouge, LA	Michael Griggs
Soil Technology, Inc.	Bainbridge Island, WA	Harold Benny
STS Consultants	Northbrook, IL	William Quinn
University of New Haven	New Haven, CT	Gregory P. Broderick
University of Tennessee	Knoxville, TN	Eric C. Drumm
University of Texas	Austin, TX	John J. Bowders
University of Wisconsin	Madison, WI	Craig H. Benson
Vector Engineering, Inc.	Littleton, CO	Kenneth R. Criley
Woodward-Clyde Consult.	Totowa, NJ	Gregory Thomas

The GCL used for the testing program was a geotextile-encased, needle-punched GCL. The method of testing was GRI Method GCL-2 (a copy of this method is available from the Geosynthetic Research Institute, Drexel University, West Wing Rush Building, Philadelphia, PA 19104). The method is similar to ASTM D5084, except (1) the maximum effective confining stress is set at 69 kPa; (2) the backpressure is set at 276 kPa; (3) the pressure gradient across the test specimen is set at 35 kPa; and (4) the permeant liquid is distilled water. Each participant was provided with a copy of the testing procedure. A sample measuring roughly 300 mm by 300 mm was cut from the same roll and shipped via overnight express to each laboratory. The first shipment was unsuccessful because the authors neglected to take steps to seal the edges of the samples; excessive bentonite fell out from between the geotextiles during shipment. New samples were cut, the edges were taped to seal the bentonite between the geotextiles, and the specimens were again shipped via overnight express. Each laboratory was asked to perform 3 tests to assist with evaluating intra-laboratory variability.

In order to establish a baseline with which to compare the round-robin test results, the GCL manufacturer provided the last 7 months of internal quality control (QC) data for this GCL product. The testing method used by the manufacturer was the same as for the round-robin program, GRI Method GCL-2. The variability of the QC testing results over time is shown on Fig. 5, while a histogram of the data is shown on Fig. 6. As indicated by the bell-shaped distribution of the logarithm of hydraulic conductivity (Fig. 6), a log-normal distribution model provides a reasonable fit of the measured data. The median (or geometric mean) of the measured data, which is a measure of the central tendency of the hydraulic conductivity values, is 6.0×10^{-10} cm/s. The coefficient of variation (the standard deviation divided by the arithmetic mean), which indicates the magnitude of variability in the data with time, is 35 percent.

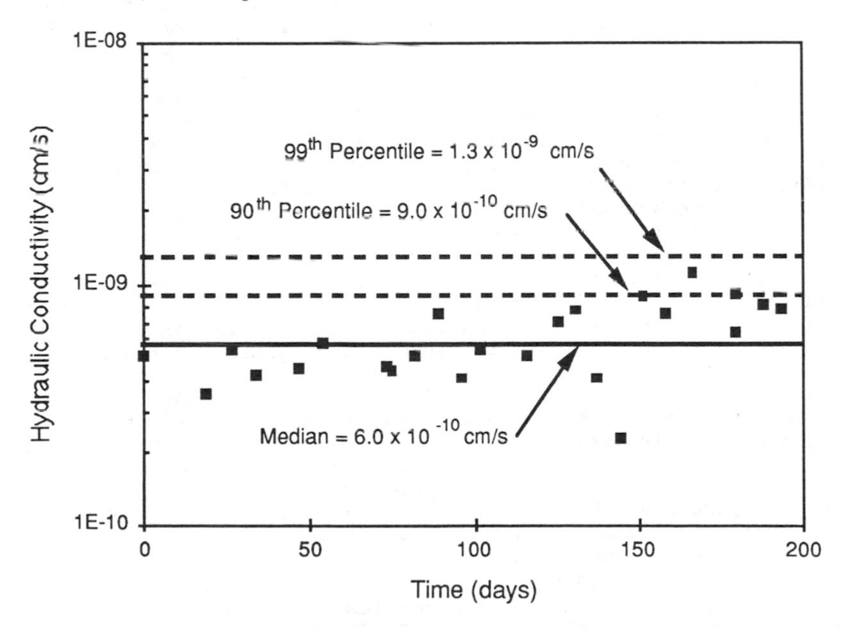

FIGURE 5 -- Hydraulic conductivity measured on internal QC purposes over a 7-month period (data courtesy of GCL manufacturer).

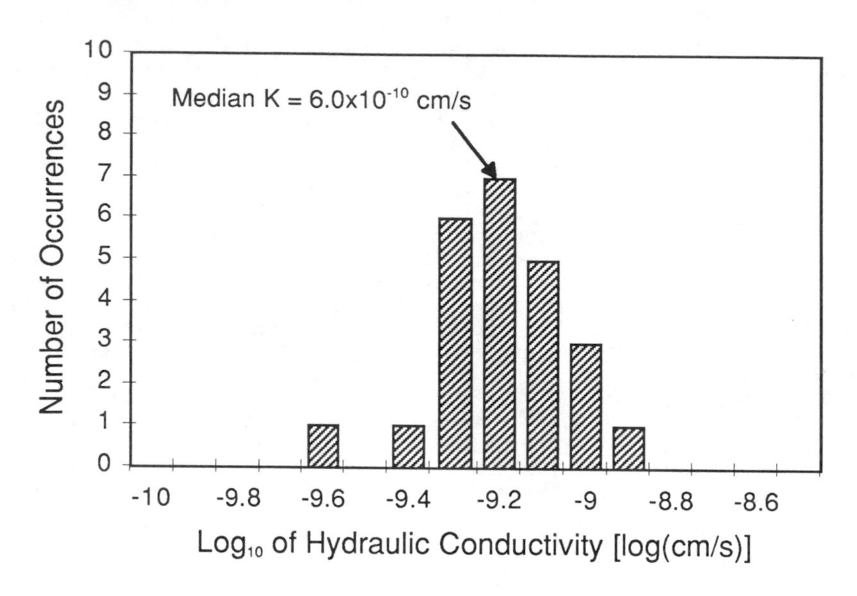

FIGURE 6 -- Histogram of manufacturer's QC data.

The variability in the hydraulic conductivity data with time could result from two sources: variability in the material and testing error. Successive test results are not strongly correlated (Fig. 5), indicating either that the variability is due to random testing error or that the material properties vary randomly over approximate 7-day intervals. The effect of variability, regardless of its source, is reflected in the 90th and 99th percentile values shown in Fig. 5 for the measured hydraulic conductivity. The 90th percentile value of 9.0×10^{-10} cm/s indicates that 1 in 10 specimens on average will have a measured hydraulic conductivity that exceeds 9.0×10^{-10} cm/s. Similarly, the 99th percentile value of 1.3×10^{-9} cm/s indicates that 1 in 100 specimens on average will have a hydraulic conductivity measurement that exceeds 1.3×10^{-9} cm/s. These thresholds can be used to identify statistically significant deviations in product quality. For example, several successive test results with hydraulic conductivity values exceeding 1.3×10^{-9} cm/s should be investigated because they are not likely the result of random variability's in material or testing.

The round-robin results are summarized in Table 2 and shown on Fig. 7. In this table the experience level of the laboratory is characterized based on a purely subjective assessment by the authors. Note that the order of the laboratories listed is different from Table 1 for anonymity. Also, several laboratories did not successfully complete all 3 tests. The median value for the round-robin results is 7.5×10^{-10} cm/s, while the coefficient of variation is 42 percent. For comparison purposes, the median, 90th percentile and 99th percentile values from the QC data are included on Fig. 7.

The median of the round-robin results is greater than that for the QC data (a median of 7.5×10^{-10} versus 6.0×10^{-10} cm/s). In addition, the variability in the round-robin results is greater than that in the QC data (a coefficient of variation of 42 versus 35 percent). These differences are reflected on Fig. 7: nearly one-third of the round-

TABLE 2 -- Summary of round-robin test results.

Lab	Experience*	Hydraulic Conductivity (cm/s)			
		Specimen 1	Specimen 2	Specimen 3	Median
A	2	7.20E-10	4.60E-10	6.40E-10	6.07E-10
B	3	6.20E-10	6.30E-10	6.20E-10	6.23E-10
C	1	1.42E-09	1.56E-09	2.12E-09	1.70E-09
D	3	5.00E-10	3.80E-10	2.40E-10	3.73E-10
E	1	1.40E-09	1.00E-09	5.60E-10	9.87E-10
F	3	7.50E-10	1.15E-09	1.10E-09	1.00E-09
G	1	6.07E-10	5.98E-10	5.62E-10	5.89E-10
H	1	1.15E-09	9.20E-10		1.04E-09
I	2	7.17E-10	6.40E-10	6.96E-10	6.84E-10
J	2	9.60E-10	1.10E-09	9.60E-10	1.01E-09
K	1	6.00E-10	5.50E-10	8.00E-10	6.50E-10
L	1	7.00E-10			7.00E-10
M	2	8.10E-10	6.50E-10	6.80E-10	7.13E-10
N	1	1.30E-09			1.30E-09
O	3	1.00E-09	9.70E-10	6.81E-10	8.84E-10
P	3	5.90E-10	6.20E-10	5.90E-10	6.00E-10
Q	3	5.81E-10	4.54E-10	6.56E-10	5.64E-10
R	1	9.00E-10	1.20E-09	7.60E-10	1.05E-09

* 1 indicates an inexperienced lab, 2 a moderately experienced lab, and 3 a very experienced lab in testing GCLs for hydraulic conductivity.

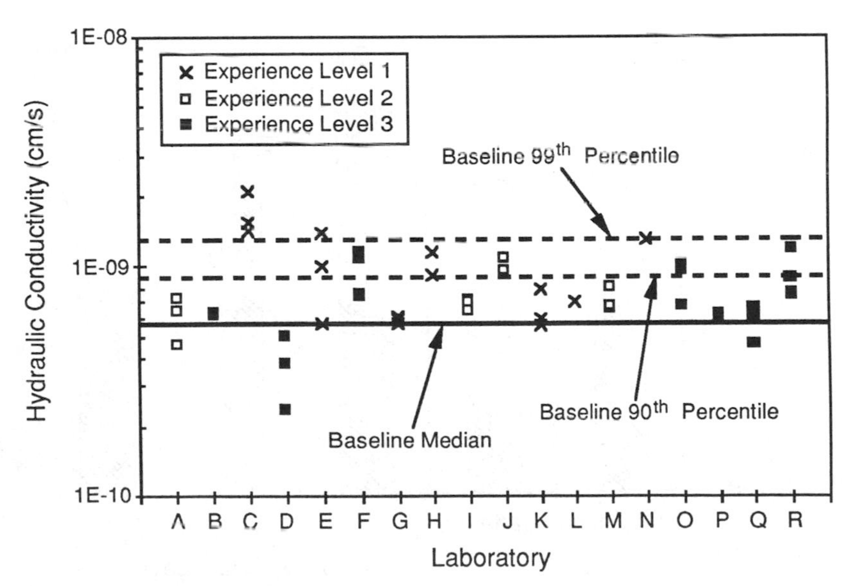

FIGURE 7 -- Summary of round-robin results (90th and 99th percentile values shown in plot are from QC data, Fig. 5).

robin test results exceed the 90th percentile from the QC data, and ten percent of the round-robin results exceed the 99th percentile value from the QC data.

The round-robin test results were separated into different categories to better understand the sources of variability in the data. First, the laboratories were divided into two categories: industry versus academic. The median values for these two categories were nearly identical, as shown on Fig. 8. Second, the laboratories were divided into three categories based on their level of experience in testing GCLs (Table 2). The level of experience clearly affected the test results (Fig. 8). The median hydraulic conductivity for the least experienced laboratories (9×10^{-10} cm/s) was about 50 percent greater than that the median hydraulic conductivity (6×10^{-10} cm/s) for the most experienced laboratories. This effect of testing experience is also shown on Fig. 7; all of the round-robin test results that exceeded the 99th percentile value from the QC data were associated with the least experienced laboratories. The influence that testing experience has on the measured hydraulic conductivity could possibly be related to specimen preparation. Trimming the test specimen is perhaps the most difficult part of measuring the hydraulic conductivity for a geotextile-encased GCL. Loss of bentonite along the edges may account for the increase in measured hydraulic conductivity with decreasing testing experience.

The round-robin test results for the most experienced laboratories (Experience Level 3 in Table 2) are comparable to the QC test data. The median for Experience Level 3 is 6.3×10^{-10} cm/s, while the median for the QC results is 6.0×10^{-10} cm/s. Similarly, the coefficients of variation are 35 and 36 percent, respectively, for the Experience Level 3 laboratories and the QC data. Therefore, the manufacturer's laboratory would seemingly fit within the Experience Level 3 category.

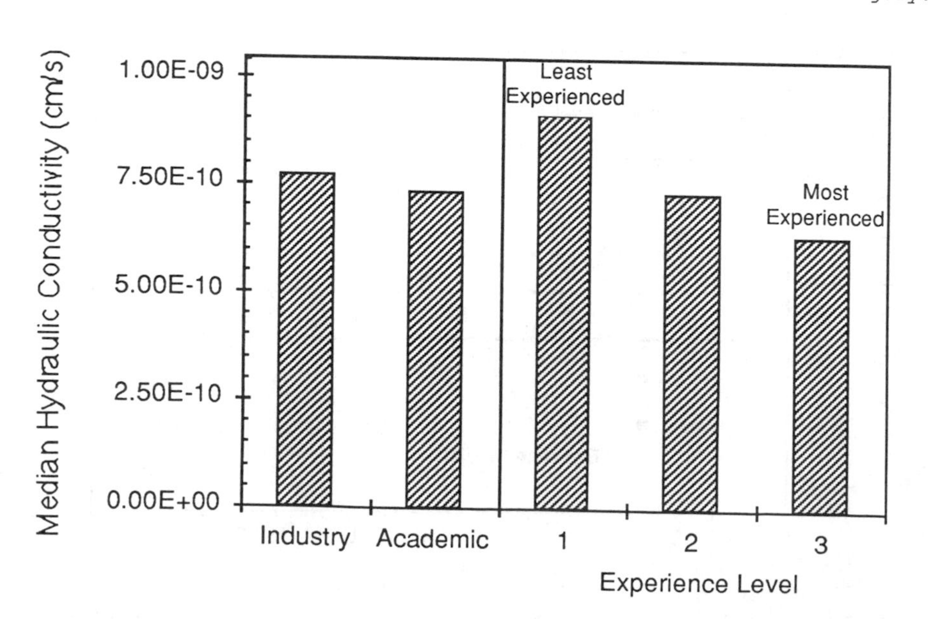

FIGURE 8 -- Effect of type of lab and experience of lab.

An analysis of the variance in the round-robin test results for Experience Level 3 laboratories indicates that the variability in test results is dominated by inter-laboratory variability versus intra-laboratory variability. Quantitatively, nearly 80 percent of the observed variability is attributed to variability in test results between laboratories. This result is significant for two reasons. First, the three laboratory specimens tested in each round-robin laboratory were obtained from the same sample to minimize the effects of material variability. Therefore, intra-laboratory variability is attributed primarily to random testing error. Since this source of variability is small, it can be concluded that random testing error is not a large source of variability in the test results (accounting for at most 20 percent). Second, there are two potential sources of inter-laboratory variability: material variability (different samples were sent to different laboratories) and systematic testing errors between laboratories. Since the coefficient of variation for the round-robin test results for highly experienced laboratories is nearly identical to that for the QC test results (all measured in one laboratory), most of the inter-laboratory variability can seemingly be attributed to material variability. The other possibility is that the round-robin sample variability is smaller than the QC data because the samples came from the same roll. In this case the inter-laboratory variability would be attributed more to systematic variations between labs. If systematic testing errors were also present (as with the less experienced laboratories), then the coefficient of variation for the round-robin results from Experience 3 laboratories would be larger than that for the QC results.

Finally, an important practical question is what value of hydraulic conductivity a design engineer should specify for a GCL. Typically, a bid specification will state a maximum hydraulic conductivity and a corresponding method of testing. A logical approach would be to specify a hydraulic conductivity that has a very low probability of being exceeded as a result of random testing variability but which has a high probability of being exceeded if the GCL is defective. On a statistical basis, 1 in 100 hydraulic conductivity's exceeded 1.3×10^{-9} cm/s in the manufacturer's QC tests and in the round-robin tests involving experienced laboratories. On a statistical basis, 1 in 100 hydraulic conductivity's exceeded 2.0×10^{-9} cm/s in the round-robin tests. Based on the results presented in this paper, a specified maximum hydraulic conductivity on the order of 2.0×10^{-9} cm/s is suggested for this particular GCL and hydraulic conductivity testing procedure. A similar process is suggested for other GCLs or testing methods.

In summary, the round-robin test results are encouraging. First, random testing errors within a single laboratory and systematic testing errors between laboratories are not dominant factors for experienced laboratories. Second, the magnitude of variability, accounting for all sources of variability including material variability, is not large for this GCL product with a coefficient of variation of approximately 35 percent for experienced laboratories. Third, an experienced laboratory should consistently (about 99 percent of the time) obtain hydraulic conductivity values less than approximately 1.3×10^{-9} cm/s for this GCL product. Finally, this upper bound should be increased only to about 2.0×10^{-9} cm/s if the experience level of the laboratory is not known.

CONCLUSIONS

Flexible-wall permeameters are routinely used for measuring the hydraulic conductivity of GCLs. The testing apparatus and procedures have been described, and the results of a round-robin testing program on a needle-punched GCL were presented. The available information indicates that the hydraulic conductivity of a GCL varies from test to test, partly as a result of variability in the material and partly as a result of variability in testing. The range of measured hydraulic conductivity, including both types of variability, is located within a single order of magnitude (2×10^{-10} to 2×10^{-9} cm/s) for the testing conditions used in the round-robin program. The coefficient of variation determined from the round-robin testing program (including experienced and inexperienced laboratories) was 42%. The only significant factor that seemed to influence the results from various laboratories was the level of experience: the highest hydraulic conductivity's were measured by inexperienced laboratories (median hydraulic conductivity $\approx 9 \times 10^{-10}$ cm/s for the least experienced labs). The coefficient of variation for hydraulic conductivity's measured by experienced laboratories was 35% (median hydraulic conductivity $\approx 6 \times 10^{-10}$ cm/s). The tendency for inexperienced laboratories to determine higher values of hydraulic conductivity is probably related to problems with proper trimming of test specimens.

Based on the results of the round-robin testing program, if a laboratory regularly (1 out of 10 times) measures a hydraulic conductivity greater than 9.0×10^{-10} cm/s on this particular GCL, using GRI Method GCL-2, they should carefully review their testing procedures. For a highly experienced laboratory, if the measured hydraulic conductivity exceeds approximately 1.3×10^{-9} cm/s, the GCL product itself should be critically evaluated.

The most important variables in hydraulic conductivity testing of GCLs include: (i) trimming the GCL specimen; (ii) determining the thickness of the specimen; (iii) selecting the effective stress; (iv) selecting the hydraulic gradient; and (v) selecting the first wetting liquid and permeant liquid. While some testing standards specify how these variables are to be handled (for instance, ASTM D5887, an index flux test, is specific on all these variables), others (such as ASTM D5084, the hydraulic conductivity standard for flexible wall permeameters) leave the decisions about these variables up to the individual who requests the hydraulic conductivity test.

ACKNOWLEDGMENTS

The participants in the round-robin program (listed in Table 1) were essential to the preparation of this paper. Their assistance is gratefully acknowledged. Mr. Robert Trauger of Colloid Environmental Technology Co. supplied the GCL material for the round-robin testing program and supplied QC testing data for the GCL; Mr. Trauger's assistance is gratefully acknowledged.

REFERENCES

1. Olson, R.E., and Daniel, D.E., 1981, "Measurement of the Hydraulic Conductivity of Fine-Grained Soils," Permeability and Groundwater Contaminant Transport, ASTM STP 746, T.F. Zimmie and C.O. Riggs, Eds., American Society for Testing and Materials, Philadelphia, 18-64.

2. Daniel, D.E., 1993, "State-of-the-Art: Laboratory Hydraulic Conductivity Tests for Saturated Soils," Hydraulic Conductivity and Waste Contaminant Transport in Soils, ASTM STP 1142, David E. Daniel and Stephen J. Trautwein, Eds., American Society for Testing and Materials, Philadelphia, 30-78.

3. Estornell, P.M., and Daniel, D.E., 1992, "Hydraulic Conductivity of Three Geosynthetic Clay Liners," Journal of Geotechnical Engineering, 118(10): 1592-1606.

4. Zimmie, T.F., 1981, "Geotechnical Testing Considerations in the Determination of Laboratory Permeability for Hazardous Waste Disposal Siting," Hazardous Solid Waste Testing: First Conference, ASTM STP 760, R.A. Conway and B.C. Malloy, Eds., American Society for Testing and Materials, Philadelphia, 293-304.

5. Daniel, D.E., Trautwein, S.J., Boynton, S.S., and Foreman, D.E., 1984, "Permeability Testing with Flexible-Wall Permeameters," Geotechnical Testing Journal, 7(3): 113-122.

6. Carpenter, G.W., and R.W. Stephenson, 1986, "Permeability Testing in the Triaxial Cell," Geotechnical Testing Journal, 9(1): 3-9.

7. Evans, J.C., and Fang, H.Y., 1986, "Triaxial Equipment for Permeability Testing with Hazardous and Toxic Permeants," Geotechnical Testing Journal, 9(3): 126-132.

8. Skempton, A.W., 1954, "The Pore Pressure Coefficients A and B," Geotechnique, 4:143-147.

9. Daniel, D.E., 1989, "A Note on Falling-Headwater, Rising Tailwater Permeability Tests," Geotechnical Testing Journal, 12(4): 308-310.

10. Jahangir, M.A., 1994, "Containment of Petroleum Hydrocarbons by Geosynthetic Clay Liners Using Natural Soil Moisture," M.S. Thesis, Univ. of Texas, Austin, TX, 200 p.

11. Bowders, J.S., 1988, Discussion of "Termination Criteria for Clay Permeability Testing," Journal of Geotechnical Engineering, 114(8):947-949.

12. Shackelford, C.D., 1994, "Waste-Soil Interactions that Alter Hydraulic Conductivity," Hydraulic Conductivity and Waste Contaminant Transport in Soils, ASTM STP 1142, David E. Daniel and Stephen J. Trautwein, Eds., American Society for Testing and Materials, Philadelphia, 111-168.

13. Ruhl, J.L., 1994, "Effects of Leachates on the Hydraulic Conductivity of Geosynthetic Clay Liners, M.S. Thesis, Univ. of Texas, Austin, TX, 227 p.

14. Shan, H.Y., and D. E. Daniel, 1993, "Results of Laboratory Tests on a Geotextile/Bentonite Liner Material," Proceedings, Geosynthetics '91, Industrial Fabrics Association International, St. Paul, Minnesota, 2: 517-535.

15. Daniel, D.E., Shan, H.Y., and J.D. Anderson, 1993, "Effects of Partial Wetting on the Performance of the Bentonite Component of a Geosynthetic Clay Liner," Proceedings, Geosynthetics '93, Industrial Fabrics Association International, St. Paul, Minnesota, 3: 1483-1496.

Specifications for GCL Applications

John W. Cowland[1]

A DESIGN PERSPECTIVE ON SHEAR STRENGTH TESTING OF GEOSYNTHETIC CLAY LINERS

REFERENCE: Cowland, John W. "**A Design Perspective on Shear Strength Testing of Geosynthetic Clay Liners,**" *Testing and Acceptance Criteria for Geosynthetic Clay Liners, ASTM STP 1308,* Larry W. Well, Ed., American Society for Testing and Materials, 1997.

ABSTRACT: Geosynthetic Clay Liners (GCLs) are being used on increasingly steep and high landfill slopes, which requires careful determination of the appropriate shear strength to be used in design. This paper reviews shear strength testing of GCLs, and the subsequent use of the strength data, from a design viewpoint.

KEYWORDS: Geosynthetic Clay Liners, Steep Slopes, Design, Shear Strength.

Geosynthetic Clay Liners (GCLs) are gaining acceptance as a replacement for clay liners in many landfills around the world. GCLs have been widely adopted as the secondary layer in composite liners for landfills in Hong Kong, where they are also currently being considered for use in landfill caps.

Hong Kong has a mountainous terrain, with many steep sided valleys, or canyons, which have 25° to 40° natural side slopes rising from near sea-level to a height of a few hundred metres. Three new landfills are under construction in these valleys, and the liners are being placed on some very steep and high slopes. In addition, new caps will soon be placed on the sloping surfaces of some old closed landfills. With highrise residential and commercial building development at the toe of many of these landfills, the stability of the landfill is an important issue.

Geosynthetic lining systems can have relatively low interface shear strengths, so they have the potential to act as a slip planes, and often form the weakest link in stability considerations. The naturally low shear strength of the bentonite component of GCLs has required careful measurements of their strength, with due consideration to the use of the measurements in design.

---

[1]Geotechnical Engineering Office, Civil Engineering Department, Hong Kong Government, 101 Princess Margaret Road, Ho Man Tin, Kowloon, Hong Kong.

MATERIAL TYPES

A typical Hong Kong landfill liner comprises a geomembrane overlying a geosynthetic clay liner (GCL). A groundwater drainage layer is usually placed below the liner, and a leachate drainage layer is placed above. Depending on various design considerations, these drainage layers may comprise crushed granitic rocks (with a geotextile cushion to protect the liner) or geonets (a planar geosynthetic material consisting of integrally connected polymeric ribs with holes to allow drainage of liquids). A typical arrangement is shown in Figure 1.

The geomembrane is usually a 2mm thick sheet of high density polyethylene (HDPE), with either smooth or roughened surfaces. On the relatively flat base areas of the landfills the geomembrane is usually roughened, or textured, on both sides. On the slopes, the geomembrane is usually textured on the side in contact with the GCL and smooth on the other; although geomembranes that are smooth on both sides are also being used.

A variety of different types of GCL are available. Some of these are shown in Figure 2. Usually, bentonite in powder or granular form is sandwiched between two geotextiles; although in one type of GCL the bentonite is attached to a geomembrane. (This latter type has not yet been used in Hong Kong). Sometimes the bentonite is mixed with a certain proportion of glue, or adhesive, to help contain it within the geotextiles, or attach it to the geomembrane. In some types of GCL, stitching is used to bind the layers together, and in other types fibres from one of the geotextiles are needle punched through to the other geotextile. Sometimes, improved anchoring of these needle punched fibres to the geotextile is accomplished through a heat burnishing process.

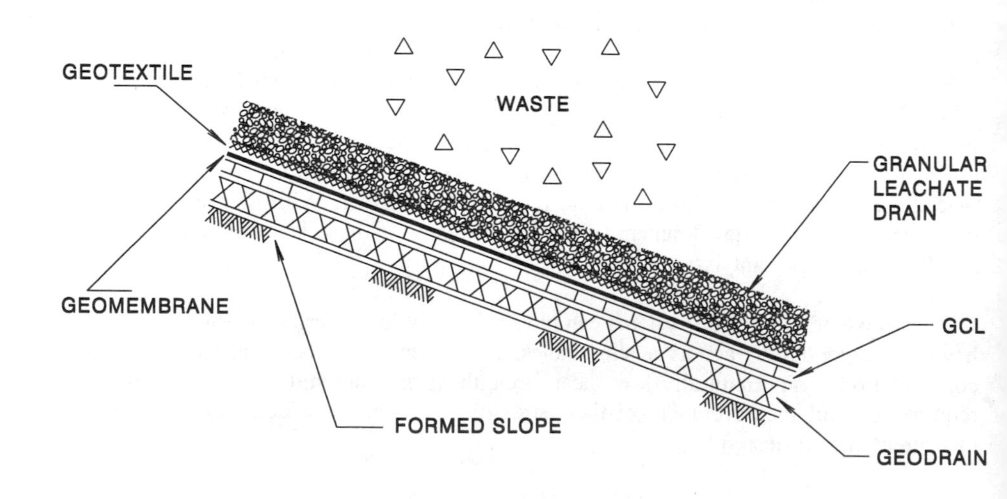

Figure 1 - Typical Hong Kong Lining System

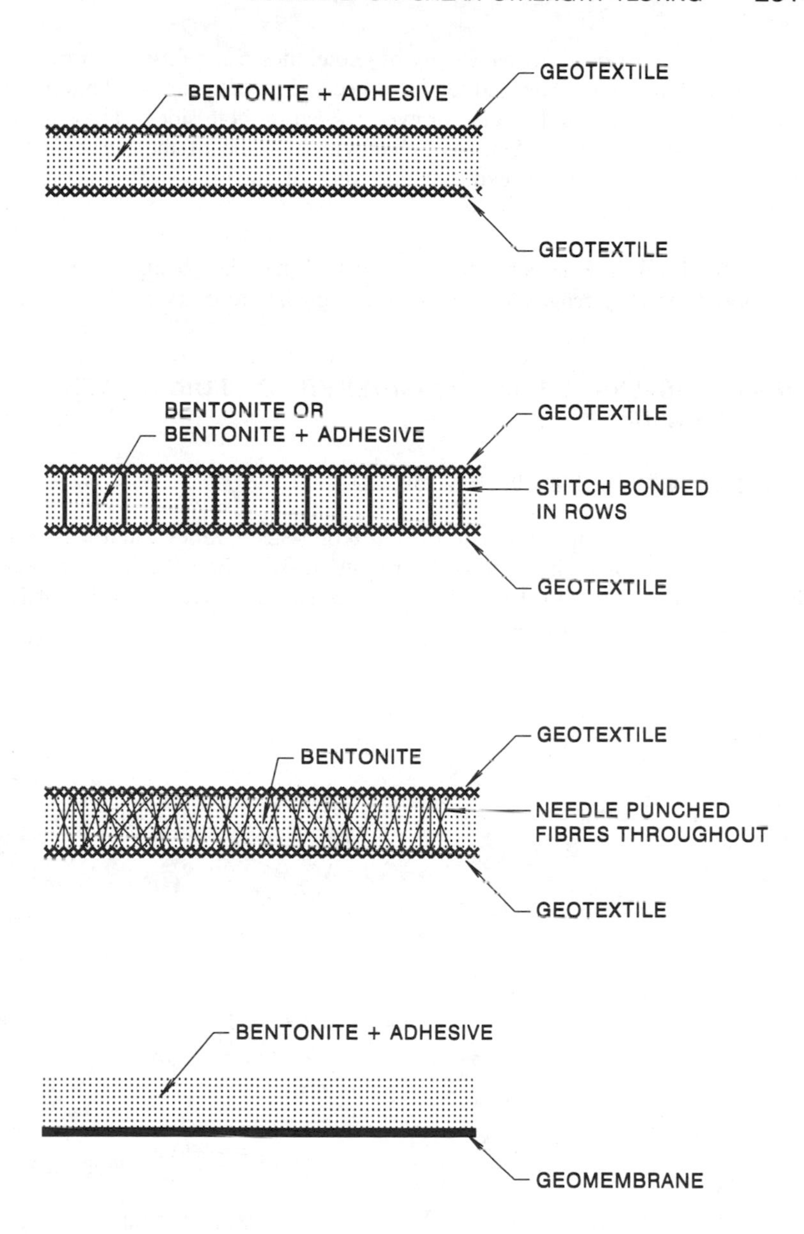

Figure 2 - Some Currently Available Geosynthetic Clay Liners (GCLs)

An almost bewildering variety of geotextiles can be used to form the top and bottom of a GCL. Non-woven geotextiles have been used both top and bottom, or non-woven on one side and woven on the other, or even woven on both sides. These geotextiles are made from polyester, polypropylene or polyethylene. One of the latest GCLs on the market incorporates a woven geotextile within one of the non-woven geotextiles, in order to strengthen the product.

With all these possible combinations of materials being placed on steep slopes, the appropriate shear strength to be used in design has to be determined carefully.

DESIGN ISSUES TO BE CONSIDERED IN FORMULATION OF TESTING PROGRAMME

Maximum and Minimum Induced Stresses

A typical section through a Hong Kong steep valley landfill is shown in Figure 3. With the use of GCLs in the basal liner system, from the bottom to the top of the side slope liner system, and within the landfill cap, the stresses induced in the GCL can range from a very high value to nearly zero.

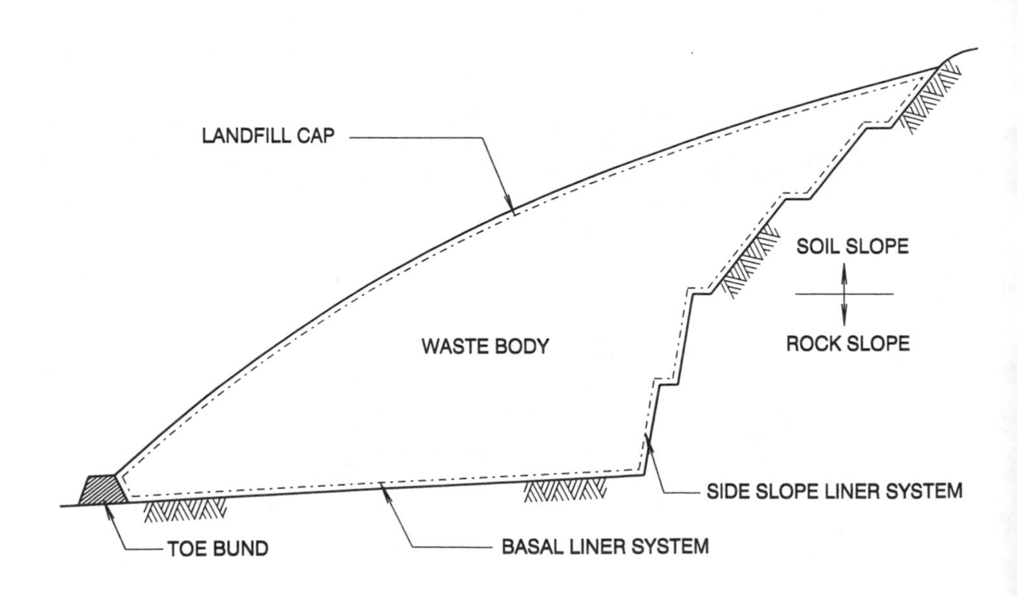

Figure 3 - Typical Section Through a Steep Valley Landfill

The maximum depth of waste planned for Hong Kong's landfills is currently around 150 metres. With a possible unit weight of 1.4 tonnes/cubic metre [1], this depth of waste could impose a stress of around 2000 kPa on the GCL. At the other extreme, the depth of soil overlying a GCL in the landfill cap may be less than one metre, giving rise to an induced stress as low as 10 kPa. In between, the stresses on a GCL on a side slope will vary from the top to the bottom of the slope. Thus, the normal stresses used in the GCL testing programme may need to range from 10 to 2000 kPa.

When planning the testing programme, all possible loading cases need to be considered. For instance, although it may be intended to place a great depth of waste over a GCL, it is still possible for a slip failure to occur during construction of the liner, caused by the downdrag of an overlying granular leachate drainage or protection layer, even though that layer may only be inducing a low stress.

Degree of Hydration

The strength of a GCL usually decreases as it becomes wet. Although the overlying geomembrane will help to keep it dry, with the typical lining system there is a groundwater drain immediately below. Experience has shown that, with its high affinity for water, the bentonite component of the GCL will soon start to become hydrated even when only a small amount of moisture is present in the underlying drainage medium.

Some designers try to keep the GCL dry by encasing it between two geomembranes. However, the long term effectiveness of this measure is unknown. A small number of defects may be inadvertently placed in the geomembranes during construction [2, 3 & 4]. Settlement and decomposition of the waste could cause movements which could possibly damage the liner, and it is also possible that the geomembranes could deteriorate during the life of the landfill.

Close examination of liners during construction in Hong Kong has produced salutory warnings with regard to the extent of hydration. In two instances, which led to exhumation of the GCL, it was noted that heavy rainfall managed to penetrate through small imperfections in the geomembrane. The surprising finding of the exhumations was the extent of lateral moisture travel along the GCL/geomembrane interface. Even with an overlying load, it was observed that water easily travels through the squashed remains of wrinkles in the geomembrane to wet a large surface area of the underlying GCL.

Clearly, the extent of hydration in the field is an important subject needing to be researched. In the meantime, it is prudent for the landfill designer to require GCLs to be tested with a certain degree of hydration.

Relevant Shear Surface

The bentonite layer within a GCL can have a low shear strength, especially when hydrated, resulting in a low internal shear strength through the mid-plane of the GCL. In addition, the bentonite can be squeezed through the opening of the geotextiles, producing a low friction surface at the interface between the GCL and the adjacent material.

Exhumation of a construction failure in Hong Kong has shown that slip surfaces can pass through the mid-plane of the GCL, and also along interfaces between the GCL and adjacent layers. Thus, the testing should be arranged to examine both these potential failure modes.

Creep and Durability

Although some research testing is underway, very little information has been published on the creep characteristics of GCLs. The issue of whether creep is likely to occur in a GCL is complicated by the determination of the amount of long term load that will be transmitted to the stitching or needle punched fibres. However, it is worth noting that the onset of creep behaviour of geosynthetics is heavily dependent on temperature, and the temperatures measured in Hong Kong landfills are relatively high. These temperatures have varied from 30°C. in waste that was a few years old to 60°C. in fresh waste.

GCLs have not yet been extensively tested for durability. Geotextiles recovered after years of burial in soil often show little degradation, and the needle punched or stitched fibres should be relatively well protected by the bentonite. However, some testing is still needed of the durability of GCLs in contact with leachate, or the vapours beneath a landfill cap; at least in research laboratories.

TESTING

Peel Tests

The peel test is a useful quality control test [5]. In the peel test, the top and bottom geotextiles are gripped and pulled apart. The force needed to pull the material apart gives a useful indication of strength consistency, and it can be related to more accurate determinations of strength from direct shear testing. This is a useful index test, which can be used for quality control both during manufacturing and during deployment in the field. However, the peel test cannot replace shear box testing for a careful evaluation of the data needed for design.

Direct Shear Tests

Direct shear tests, with careful measurements of loads and displacements, allow a considerable amount of data to be obtained on the shear strength behaviour of a GCL. The ASTM Test Method for Determining the Coefficient of Soil and Geosynthetic, or Geosynthetic and Geosynthetic, Friction by the Direct Shear Method (D 5321-92) gives guidance on how to carry out the test. In addition to the details contained in that testing standard, the designer may wish to consider the following aspects.

Degree and Method of Hydration--The current practice is for the GCL to be placed in the shear box, with a suitably representative normal load placed on top, and then for the box to be immersed in water for 24 hours before carrying out the shear test. This is intended to produce a reasonable replication of the wetting that may occur in the landfill. However, observations of worst case wetting from field exhumations in Hong Kong (as discussed above) suggest that it would be prudent to consider a smaller normal load and a longer immersion period.

In addition, basic data to be recorded during the test should include the initial moisture content of the GCL, and its moisture content after testing.

Constraint of Failure Surface--It is interesting to consider whether, with a multilayered system, the whole system should be tested at once. However, it is far easier to analyse the results of testing only one interface at a time. In order to constrain the failure surface to test one particular GCL interface or shear plane, a rigid substrate is needed with some form of textured gripping surface which will provide an even grip over one surface of the GCL.

Strain Rate--In the shear testing of clays, there is a need to keep the strain rate as low as possible in order to allow excess pore water pressures to dissipate to obtain drained, rather than undrained, shear strength parameters. With the low permeability of bentonite, the strain rate needs to be especially low. For instance, it was found in one programme of GCL interface shear testing that the friction angles measured at a strain rate of 0.04mm/min were 1-2° lower than at a strain rate of 1.0mm/min.

Plan Area of Shear Box--The shear box should be as large as possible, with a minimum size of 300 mm. The nature of geosynthetic samples allows the travelling half of the shear box to have a larger plan area than the fixed half, so that the area correction usually used for soils is not needed.

Extent of Testing

The determination of GCL shear strength needs to be both material specific and project specific. With the rapid development of GCLs, incorporating different geotextiles and methods of bonding them together, it is not possible to rely on previous test results. The designer needs to know the shear strength characteristics of the individual GCL to be used in his landfill.

The testing programme for each particular landfill needs to encompass the expected physical conditions (eg. loads) and all the interfaces between each of the materials used in the leachate lining and drainage system. A series of direct shear tests will need to be carried out during the detailed design of the landfill, in order to obtain stability design data. Conformance tests will then need to be performed on the materials delivered to site. It is normally adequate to use index tests to determine material conformance.

Independent Testing

Many manufacturers have sophisticated laboratories where good quality testing is carried out. However, the landfill designer needs to be assured that the strength data is independent of any commercial considerations. This is best arranged through testing in a reputable independent laboratory.

USE OF STRENGTH DATA IN DESIGN

Having obtained strength data from careful testing, the designer needs to consider various aspects in the use of this data in design.

Peak or Residual Strength

With increasing strain, GCLs typically exhibit a strength that rises to a peak and then falls to a residual value. The difference between the peak strength and the residual (or stabilized post peak) strength is relatively small in some products, but can be quite large in other products. Although it varies from one type of GCL to another, and with the amount of normal stress, the peak strength is usually achieved after 20 - 30 mm of displacement.

Even for a stable landfill, it is to be expected that some parts of the liner may experience more than 20 - 30 mm of displacement. Domestic waste, for instance, is a relatively compressible material, and settlements may give rise to quite large down-drag forces on the side slope liners.

Our limited knowledge of landfill failures suggests that it would be unwise for the designer to rely on peak strength throughout the lining system. From a finite element back analysis of the Kettleman Hills Landfill failure, Byrne [6] found that the onset of progressive failure leads to residual strengths on the side slopes, although the peak strength may still be relied upon for a proportion of the base liner.

As noted by Koerner [7], the analysis of side slope liner stability is complicated for a multi-layered liner and leachate collection system. The unit load of the waste gravitationally induces shear stress through the leachate collection system onto the liner and then onto the groundwater collection system. Depending on the frictional characteristics of the surfaces involved, it is possible that only a portion of the induced stress is transmitted to the layer below.

Therefore, it is possible, especially when a low friction slip surface is built into the upper part of the system, that some interfaces may not progress to their residual strengths. However, field measurements, and a thorough finite element or finite difference analysis of the stresses and strains in the system [8] would seem to be needed to determine whether individual interfaces should be designed using the measured peak or residual strengths.

Deformation of the Shear Surface

With a coarse granular leachate drainage layer above the lining system (Fig. 1), and high normal loads, it is likely that the shear surface will become deformed. This deformation may result in an increase of the friction angle. However, in order to utilise this increased friction angle, the designer needs to be confident that the same deformation will occur in the laboratory and the field. The gravel used in laboratory tests needs to be truly representative of the gravel used in the landfill. Also, the normal loads in the laboratory tests should be varied to ensure the deformation would still exist with lower than expected overlying loads in the field.

Landfill Loading Sequence

Load should be placed above the GCL as soon as possible after installation, in order that it does not get the chance to swell and weaken, due to any moisture that may be available.

With a low strength lining material being placed on steep slopes, it would appear that there is a high chance of a slip occurring. Luckily, even though GCLs have an inherently low strength, a slope failure will only occur if the imposed loads push it down the slope. Giroud and Beech [9] demonstrate that many of the stability problems associated with these materials can be avoided by utilising an incremental balanced method of placing the overlying load (Figure 4).

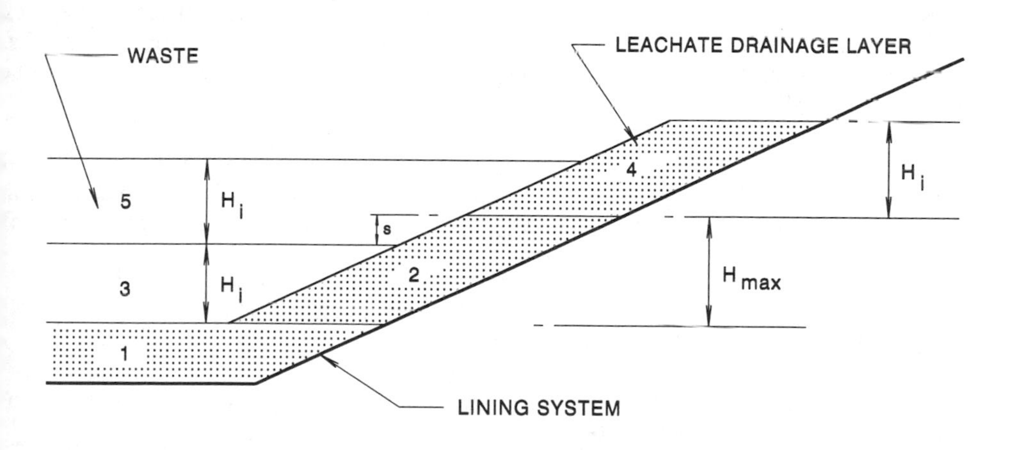

Figure 4 - Incremental Placing of Load (After Giroud & Beech)

In this method, the potential for the leachate drainage layer, or soil cover, to slide down the slope is incrementally balanced by the propping force provided by horizontal layers of waste. The leachate drainage layer, or soil cover, is placed in stages (2, 4, etc.) alternating with waste placement stages (3, 5, etc.). The maximum calculated increment height, H_{max} , beyond which the system could become unstable, is used only for the first increment. Subsequent increments, H_i , are a step, s, smaller to ensure that the lining system is always protected from waste placement operations.

This solution is viable only if the waste is properly placed and compacted to ensure stability, and the temporarily exposed geosynthetics are not adversely impacted by exposure to sunlight. Thus, it is important for the designer to carefully stipulate how the loading stages and waste layers are to be placed.

CONCLUSIONS

The geosynthetic clay liners (GCLs) currently being used in landfill liners and caps have the potential to act as slip planes, as the bentonite they contain has a low shear strength, especially when wet. For the designer, it is important to incorporate shear strengths obtained from relevant and independent testing into the design of the landfill.

The designer needs to consider the complete range of loads that may be exerted on the GCL, the possible degree of hydration, and all possible shear surfaces. For drained shear strength parameters, it is important to maintain a low rate of strain during the test.

The determination of GCL shear strength needs to be both material specific and project specific. With the rapid development of GCLs, incorporating different geotextiles and methods of bonding them together, it is not possible to rely on previous test results. The designer needs to know the shear strength characteristics of the individual GCL to be used in his landfill.

Although GCLs have an inherently low strength, a slope failure will only occur if the imposed loads push it down the slope. Many of the stability problems associated with these materials can be avoided by utilising a carefully designed incremental method of placing the overlying load.

ACKNOWLEDGEMENT

This paper is published with the permission of the Director of Civil Engineering of the Hong Kong Government.

REFERENCES

[1] Cowland J.W., Tang K.Y. and Gabay J. (1993). "Density and Strength Properties of Hong Kong Refuse", Fourth International Landfill Symposium, Sardinia, Italy, pp. 1433-1446.

[2] Bonaparte R. and Gross B.A. (1990). "Field Behavior of Double Liner Systems", ASCE Special Publication on Waste Containment Systems, pp. 52-83.

[3] Gross B.A., Bonaparte R. and Giroud J.P. (1990). "Evaluation of Flow from Landfill Leakage Detection Layers", Fourth International Conference on Geotextiles, The Hague, Vol. 2, pp. 481-486.

[4] Workman J.P. (1993). "Interpretation of Leakage Rates in Double Liner Systems", Geosynthetic Liner Systems: Innovations, Concerns and Designs, R.M. Koerner and R.F. Wilson-Fahmy, Eds., Industrial Fabrics Association International, pp. 95-112.

[5] Heerten G., Saathof F., Scheu C. and von Maubeuge K.P. (1995). "On the Long Term Shear Behaviour of Geosynthetic Clay Liners in Capping Sealing Systems", International Symposium on Geosynthetic Clay Liners, Germany, pp. 141-150.

[6] Byrne R.J. (1994). "Design Issues with Strain-Softening in Landfill Liners", Proceedings of Waste Tech '94, National Solid Wastes Management Association, South Carolina, USA, Section 4, pp. 1-26.

[7] Koerner R.M. (1994). "Designing with Geosynthetics", Prentice Hall, 783p.

[8] Long J.H., Gilbert R.B. and Daly J.J. (1994). "Geosynthetic Loads in Landfill Slopes: Displacement Compatibility", ASCE Journal of Geotechnical Engineering, Vol. 120, No. 11, pp. 2009-2025.

[9] Giroud J.P. and Beech J.F. (1989). "Stability of Soil Layers on Geosynthetic Lining Systems", Geosynthetics '89 Conference, San Diego, USA, pp. 35-46.

Kent P. von Maubeuge[1]

Manufacturing Quality Control and Specification Criteria for Geosynthetic Clay Liners

REFERENCE: von Maubeuge, K. P., **''Testing and Acceptance Criteria for Geosynthetic Clay Liners,''** *Testing and Acceptance Criteria for Geosynthetic Clay Liners, ASTM STP 1308,* Larry W. Well, Ed., American Society for Testing and Materials, 1997.

Abstract

Geosynthetic clay liner (GCLs) combine geological materials with synthetic fabrics and require a new approach to conformance and quality control testing. GCLs have been used in various sealing applications over the last 10 years and have found an increased use in landfill applications. Therefore during the June 1993 ASTM meetings, a GCL subcommittee, D 35.04 was created with Mr. Larry Well of CH2M Hill, Portland, OR. serving as subcommittee chairman. The subcommittee comprises 6 task groups:

- Physical properties
- Manufacturing quality control and assurance (MQC/MQA),
- Logistics,
- Endurance,
- Hydraulic properties, and
- Mechanical properties.

As an example, the *Manufacturing QC/QA* task group has currently developed a standard of practice for manufacturing quality control of GCLs, and methods for determining a swell index value for clay and the fluid loss of the clays. A standard for manufacturing quality assurance is currently in development.

In the past quality control and quality assurance programs were developed by various manufacturers resulting in QC/QA programs which benefit the manufacturers´ products and not necessarily the project requirements. Another problem is the fact that the construction quality control and quality assurance (CQC/CQA) plan is often written in isolation by different people from those writing the MQC plan [1]. This paper will discuss the state-of-the-art of quality control and quality assurance for GCL testing to help to maximize GCL performance.

Keywords: Geosynthetic clay liner, Quality control, Quality, Assurance, Specification criteria, Manufacturing quality control,

[1] Naue Fasertechnik GmbH & Co. KG, Wartturmstraße 1, 32312 Lübbecke, Germany

The GCL Product

GCLs (see Fig. 5) are composite products consisting of geosynthetics and bentonite clay materials. They are formed together by means of water-soluble adhesives or shear strength transforming reinforcements such as stitches (stitch bonding) or fibers (needle-punching). Geosynthetic Clay Liners have been used in critical containment applications for over a decade and their effectiveness has been well documented both in the laboratory and in the field [2]. The development and utilization of geosynthetics (in the case of GCLs, geotextiles and geomembranes) are described in detail by Koerner [3].

The Bentonite Component of GCLs

Approximately 50 to 130 million years ago ash spewed out of volcanoes and settled to the earth surface into salt and fresh water seas creating ash deposits. Over time these deposits were covered with silt, and chemical reactions took place creating sodium clay bentonite. Bentonite was named after Fort Benton, Wyoming which is where it was discovered in about 1890. The predominant mineral in bentonite is montmorillonite and it is named after a clay deposit at Montmorillon in Southern France.

Clay minerals have played an important role throughout the cultural and industrial development of mankind. For thousands of years clays have been used for making pottery, brick, and other building materials. Today there are hundreds of uses for clay minerals. Bentonite clays in particular are used in a tremendous variety of products ranging from pet litters to pharmaceuticals to GCLs [4].

Bentonite clay is expansive and exhibits the characteristics of low permeability and high absorption capacity. A high-graded bentonite has a montmorillonite content of approx. 70 percent - 95 percent. Additionally bentonite contains other minerals such as cristobalite, quartz, mica, and feldspar carbonates. The colouring of bentonites can vary considerably and it can be cream, ochre, reddish-brown or even dark blue, but in general color has no influence on the quality of the bentonite.

A single clay crystal of montmorillonite has a plate shape, is extremely small and consists of 5 - 12 single tetrahedron (silica) / octahedron (alumina) layers. The thickness is approx. 1×10^{-9} m, the plate size varies from 2.5×10^{-9} m to 2.5×10^{-8} m. The distance between single plates is in the dry condition approx. 2.5×10^{-10} m and completely hydrated approx. 1×10^{-9} m.

Most clay crystals principally consist of a silica and alumina atomic sheet. On the silica sheet (tetrahedron shaped, see Fig. 1) the silicon (four valences) is surrounded by four oxygens (each with two valences) each bonding one valence to the silicon. Some of the remaining valences are bonded with adjoining silicons.

On the alumina sheet (octahedron shaped, see Fig. 2) the aluminium is surrounded by six oxygens/hydroxyls. Some remaining free valences are linked to adjacent aluminiums. For stability reasons a tetrahedron is linked to an octahedron. The montmorillonite is a clay mineral consisting of one alumina sheet bonded between two silica sheets (Fig. 3).

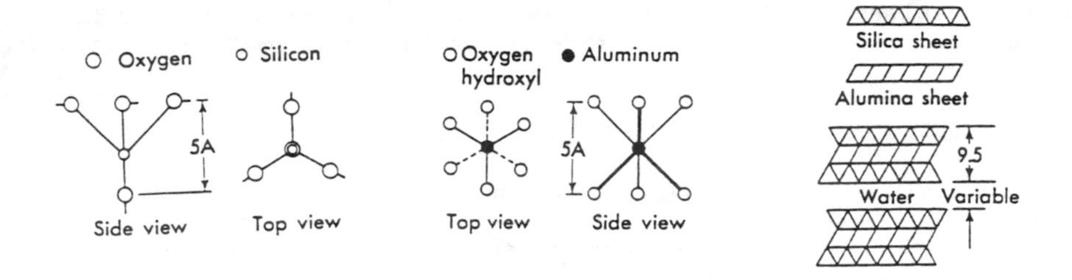

Fig. 1: Tetrahedron silica sheet [5]   Fig. 2: Octahedron alumina sheet [5]   Fig. 3: structure of montmorillonite [5]

Due to a partial substitution of silicon (Si^{4+}) through aluminium (Al^{3+}) in the tetrahedron sheet and aluminum (Al^{3+}) through a bivalent ion (e. g. Mg^{2+}) in the octahedron sheet a surface unbalance is created. This negative excess of charge is compensated by exchangeable cations (e. g. Ca^{2+}, Na^{+}), which hold the element sheets together electrostatically.

Due to the many very small plate shaped sheets the montmorillonite has a very large external and much larger internal surface (up to 800 m^2/g of clay). The negative surface, therefore, allows an adsorption of positively charged cations. This can be measured with the CEC method (cation exchange capacity, expressed in terms of the total number of positive charges adsorbed per mass per unit area). For montmorillonite a typical value of the CEC is 80 - 120 mval/100 g. Montmorillonite volume swells in contact with water approx. 900 percent. In this stage the single layers are movable against each other which results in the plasticity of the clay.

Due to the importance of these products in critical applications a consistently high quality must be assured. The specifier must be confident that the bentonite has the required sealing properties, that the geotextiles have the required strength and durability, and that the finished product is physically sTable and maintains a low hydraulic conductivity.

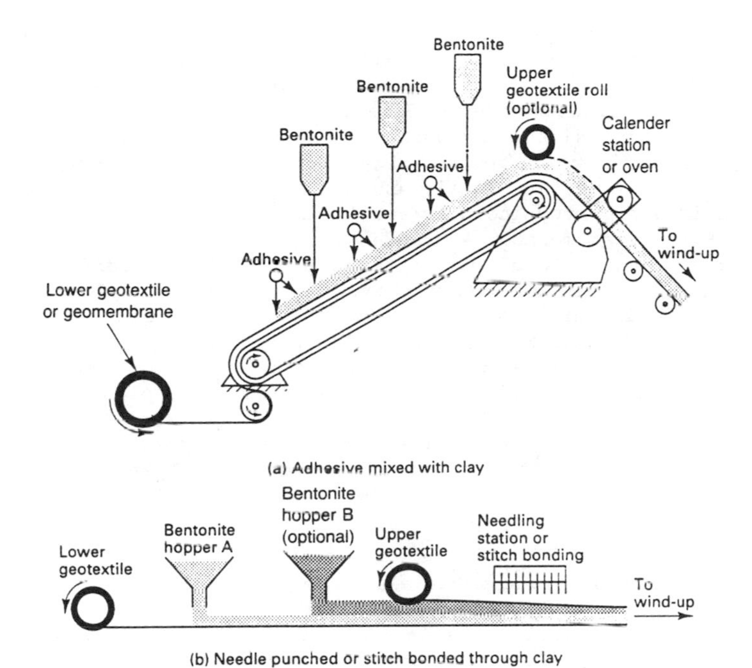

Fig. 4: Schematic GCL production processes [6]

GCLs are manufactured on continuous production lines (see Fig. 4 and 5) whereby a set quantity of natural Wyoming sodium bentonite is either confined between two geotextile layers or placed on a thin geomembrane. By means of needle-punching, stitch-bonding or glueing a composite product is held together. Adhesive bonding provides only temporary bonding and loses its strength when in contact with water. The stitch-bonding process and the needle-punching process form a longer term bonding force. The stitch bonding process bonds the two geosynthetics together in rows at a certain distance, and the needle-punching process has reinforcing fibers over the entire surface of the product.

Manufacturing Quality Control

a) QC of the Bentonite Component

Bentonite is a key component of a geosynthetic clay liner. The function of the bentonite is to maintain a low hydraulic conductivity in the hydrated state. It must possess the required water adsorption and swell potential to prevent the flow of liquids and to seal any voids that may exist. How much a bentonite swells, affects its ability to seal void spaces, its density and permeability. The increase in volume, or swell, experienced by the bentonite as it hydrates is a property known as swell index and is measured using the ASTM D5890-95 test method. A graduated cylinder is filled with 100 ml of water to which 2 g of bentonite is added. The bentonite is first milled to a powder and added slowly to the water cylinder. After leaving the cylinder undisturbed for 24 hours the volume occupied by the clay is measured and recorded. The Fluid Loss Test Method (ASTM D5891-95) is an easy-to-measure index parameter which estimates the hydraulic conductivity of bentonite supplied to the plant.

Both above mentioned test methods should be performed at a minimum frequency once per truck or rail car, or 50,000 kg. The reported value should be a minimum average value (see Table 1).

b) Geotextile/geomembrane materials

To assess the geotextile materials at the GCL production facility it is necessary to test at a frequency of least every 20,000 m² for mass per unit area and grab tensile strength in the machine direction and cross direction. It is common practice for the geosynthetic manufacturers to supply the GCL manufacturers with a certification letter of the geosynthetic components prior to arrival. These must be checked before the components are used for GCL production. The reported value for the mass per unit area is the typical value and the minimum average roll value (MARV), for the grab tensile strength the MARV (see Table 1).

Similar to geotextile quality control geomembranes should be tested for mass per unit area, thickness and tensile strength at break and yield of the machine and cross direction. The minimum frequency of testing is 20,000 m² and the report values should be the minimum average (see Table 1).

c) Finished GCL

In the past many tests appeared in GCL quality control and quality assurance programs. To clarify this situation the GCL manufacturers as well as the engineering community agreed on the following set of tests for the finished GCL:

- Clay Mass per Unit Area (dried)
- Grab Tensile Strength (machine direction and cross direction)
- Index Flux

In order to provide adequate liquid containment the bentonite must be uniformly distributed throughout the GCL. In the past a minimum bentonite mass per unit area of 4.9 kg/m² was required. On site the water content of bentonite of the needle-punched GCLs was approximately 8 percent - 15 percent whereas the adhesive bond products had a water content of approximately 20 percent - 40 percent. Naturally, it is important that the value is measured at a designated water content so that a comparison is possible. It was determined that the clay mass per unit area should be measured in dry state. Dry bentonite is defined as 0 percent water content. The frequency of testing should be a minimum of once every 4,000 m² and reported as a MARV value for the clay mass per unit area. The clay water content should be reported, if necessary, at an average value also at least every 4,000 m². It should be noted that the clay water content is for information only and should not have a project specific requirement.

The grab tensile strength of the finished product should be measured at least every 20,000 m² and should be reported as a MARV.

The effectiveness of the GCL as a hydraulic barrier is determined by measuring the index flux through the product using a triaxial cell permeameter. Even though a higher frequency of testing will give a larger degree of confidence, the frequency of testing is set at the moment at a frequency of once a week, with the last 20 values reported. The last 20 values to be reported should end at the production date of the supplied GCL. If the manufacturer has more production facilities and/or production lines, the test must be performed and reported for each line. The reported value should be the maximum value (see Table 1).

Table 1: Minimum types of tests and their frequencies for the MQC of GCLs as stated in ASTM D 5889-95 [7].

Test Designation	Test Method	Frequency of Testing	Report Value
I) Clay[1]:			
- free swell	D5890-95	1 per truck or railcar but min. every 50,000 kg	minimum average
- fluid loss	D5891-95	1 per truck or railcar but min. every 50,000 kg	minimum average
II) Geosynthetic Materials:			
(a) geotextiles			
- mass per unit area	D5261	20,000 m^2 (200,000 ft^2)	typical and MARV [6]
- grab tensile strength(MD & CD)	D4632	20,000 m^2 (200,000 ft^2)	MARV [6]
(b) geomembrane			
- mass/unit area[8]	D5261	20,000 m^2 (100,000 ft^2)	min. average [6] min. average [6]
- density	D792 or D1505	20,000 m^2 (100,000 ft^2)	min. average [6]
- thickness	D5199 or equivalent	20,000 m^2 (200,000 ft^2)	
- tensile strength at break and yield (MD & CD)	D638	20,000 m^2 (200,000 ft^2)	min. average [6]
III) Finished GCL[2]:			
- clay mass/unit area (dried)[5]		4,000 m^2 (40,000 ft^2)	MARV
- clay water content	D4643	4,000 m^2 (40,000 ft^2)	average value [7]
- grab tensile strength (MD & CD)[4]	D4632	20,000 m^2 (200,000 ft^2)	MARV
- index flux[4]	D5887-95	once weekly with the last 20 values reported [3]	maximum value

MD - machine direction; CD - cross-machine direction

Note: (1) The tests on the bentonite are to be performed on the as-received material before fabrication into the GCL product.

(2) Components from finished GCL product should not be separated and tested, because the production process may alter the properties of the components.

(3) The last 20 values to be reported should end at the production date of the supplied GCL. If the manufacturer has more production facilities and/or production lines, the tests must be performed and reported for each line.

(4) This test may not be applicable for geomembrane-backed GCLs

(5) Dried bentonite is defined as 0 percent water content.

(6) Letter of Certification from component manufacturer and/or QA from GCL manufacturer. Letters of Certification must arrive and be checked before the components are used for the GCL production. A Letter off Certification must contain manufacturer´s address, name of product, production date, lot number, roll number, required tests, required report values, signature of authorized company´s representative and his printed name.

(7) Only for information.

(8) Mass per unit area can be calculated using the density value and the thickness of the geomembrane.

Manufacturing quality control program enables the GCL manufacturer to demonstrate that the produced GCL meets the product data and is consistent in its properties. Of course, it is important that the finished product be able to withstand the rigours of construction, maintain its stability under high stress conditions and maintain a low permeability over the life of the project. Further tests for project specific requirements are sometimes needed and should be developed in cooperation with the manufacturers, the project engineer and the design engineer.

Manufacturing Quality Assurance

It is necessary to perform manufacturing quality assurance on site to ensure that the required specifications which were specified are met and the correct material is delivered. As PEGGS note: "It should not be considered to be a second chance at QC testing". [1]

Table 2: Minimum types of test and their frequencies for the MQA of GCLs on site

Test Designation	***Test Method***	***Frequency of Testing***	***Report Value***
Clay Mass/Unit Area (dried)		every 20,000 m²	average value
Grab Tensile Strength (machine and cross direction)	D 4632	every 20,000 m²	minimum average value
Index Flux	D5887-95	every 50,000 m²	maximum value

Table 2 summarizes the tests and the freqencies required for a basic quality assurance program on site. The amount of tests carried out for quality assurance need not exceed the tests listed in Table 1 which recommends the minimum quality control tests for the manufacturer of GCLs. The test frequency depends on the project size and should be carried out at least once per project, if the project size is larger than 10,000 m². Otherwise the testing frequency can be as stated in Table 2.

To ensure that the manufacturing quality control is carried out properly an independent quality assurance company should be hired to carry out manufacturing quality assurance testing.

Requiring a planned audit such as DIN ISO 9001 or a GRI accreditation can furtheron increase the confidence to the production plant.

An overview of a proposal for QC/QA from the single component all the way to the installation is shown in Table 3 and is based on PEGGS' [1] recommendation.

Specification Criteria

Typical GCL specification criteria for the QC tests listed in Table 1 are as follows:

* free swell of the clay component, 25 ml
* fluid loss of the clay component, 18 ml
* clay mass per unit area (dried), 4,000 g/m²
* index permeability of the GCL, 5×10^{-11} m/s

The overlap of installed GCLs should be at least 0.30 m and the cover material which should be installed no later than 24 hours after the GCL deployment should be no less than 0.30 m. In wet conditions the water content of GCLs should be checked to ensure that it does not exceed a value of 50 percent moisture content otherwise damage during installation can occur especially if coarser grain material is used. Traffic over hydrated but covered GCLs can cause rutting which can squeeze the bentonite sideways if the cover layer is not selected thick enough. In general, the cover thickness of areas with high traffic should be approx. 0,5 m - 1 m thick during the installation process, depending on the traffic and the cover material.

In cases where GCLs are designed for slope applications the specification for the GCL has to include the critical interface, the long-term shear resistance and the internal shear strength of the GCL. In most cases these values must be clarified prior to the design.

The puncture resistance of the geotextiles as well as the thickness and mass per unit area of the geomembranes (if involved) are also project specific parameters and need to be specified on a project-to-project basis.

Table 3 gives an overview of minimum considerations for the GCL specifications and it is necessary to discuss these in the pre-design phase with all involved people. In any case it is necessary to follow ASTM standards as listed in Table 1, 2, and 3.

Summary

The GCL manufacturers as well as the engineering community are currently working on GCL standards and in near future various standards will be published.

This paper presents the current state of manufacturing quality control standard and gives additionally an idea how a quality control and quality assurance program can look like involving the single components of the GCL up to the installation.

It is obvious that project specific designs will require additional test methods and quality assurance programs, but it is also obvious that these should then be project specific and not product specific.

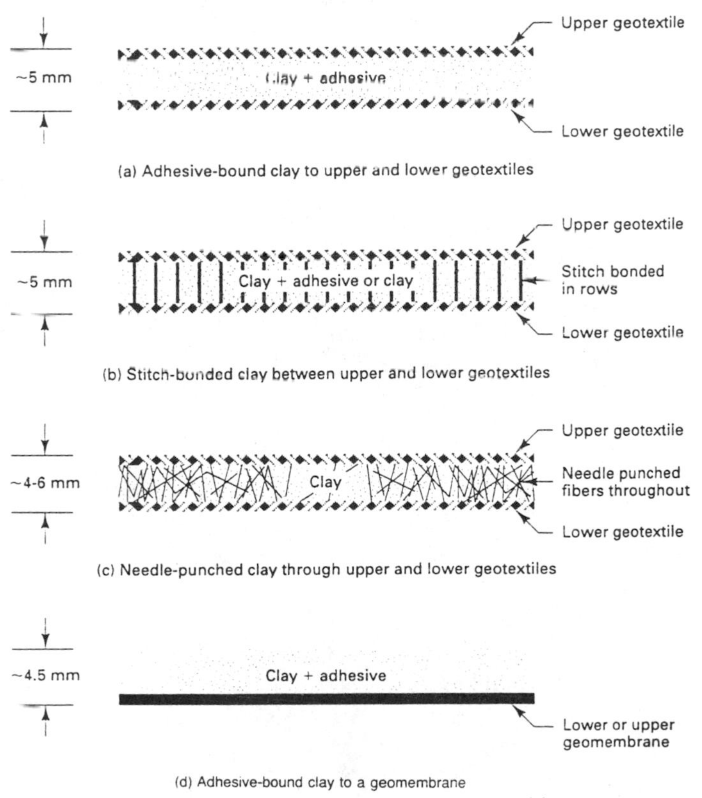

Fig. 5: Cross section of currently available GCLs [8]

References

[1] PEGGS, D. IAN and TRAUGER, BOB, "Quality-Testing Program for Geosynthetic Clay Liners", MSW Management, November/December 1995.
[2] VON MAUBEUGE, KENT. "Performance of Geosynthetic Clay Liners (GCLs)", University of Toronto, GCLs for Waste Containment and Pollution Prevention, Toronto, June 1995
[3] KOERNER, ROBERT M. "Designing with Geosynthetics", Third Edition, ISBN 0-13-847823-6, 1994.
[4] TRAUGER, R. J. "The structure, properties and analysis of bentonite in Geosynthetic Clay Liners", Proceedings of the 8th GRI Conference, December 13-14, 1994, Philadelphia, PA
[5] "Soil Mechanics and Foundations", New York 1979
[6] EPA, DANIEL, KOERNER, CARSON, "Quality Assurance and Quality Control for Waste Containment Facilities" Cincinnati, September 1993
[7] ASTM 5889-95, Standard Practice for Quality Control of Geosynthetic Clay Liners
[8] KOERNER, ROBERT M., GARTUNG, E., and ZANZINGER, H., "Geosynthetic Clay Liners", International Symposium, Nürnberg, April 1994

Patrick J. Fox[1], Daniel J. De Battista[2], and Shu-Hong Chen[2]

A STUDY OF THE CBR BEARING CAPACITY TEST FOR HYDRATED GEOSYNTHETIC CLAY LINERS

REFERENCE: Fox, P. J., De Battista, D. J., and Chen, S.-H., **"A Study of the CBR Bearing Capacity Test for Hydrated Geosynthetic Clay Liners,"** *Testing and Acceptance Criteria for Geosynthetic Clay Liners, ASTM STP 1308,* Larry W. Well, Ed., American Society for Testing and Materials, 1997.

ABSTRACT: A modified version of the California Bearing Ratio (CBR) penetration test was used to investigate the bearing capacity of hydrated geosynthetic clay liners (GCLs). The bearing capacity of GCLs is a concern because, once hydrated, the soft bentonite may squeeze laterally under concentrated loads. A series of bearing capacity tests were performed to study the influence of GCL product type, surcharge pressure, displacement rate, mold diameter, and termination criterion on lateral bentonite squeezing. Test results show that adhesive-bonded GCLs have a somewhat greater tendency for bentonite squeezing than reinforced GCLs. The magnitude of surcharge pressure during hydration is an important variable for GCL bearing capacity testing. The rate of piston displacement has little influence on the bearing capacity of hydrated GCLs. Confining molds having a diameter ≥ 235 mm are satisfactory for the measurement of GCL bearing capacity. A standard CBR mold with a diameter of 152 mm is not recommended. Termination criteria based on a specified force or a peak in the force-displacement relationship may be unreliable for some tests. A specified displacement termination criterion of 20 mm was found to be more appropriate for the GCL bearing capacity tests conducted in this investigation.

KEYWORDS: geosynthetic clay liner, bentonite, bearing capacity, cover soil, CBR test

---

The bearing capacity of hydrated geosynthetic clay liners (GCLs) is a concern after installation because the bentonite may squeeze laterally under concentrated loads. Thus, the thickness of the cover soil should be such that the underlying GCL is protected from stress concentrations due to construction activities and permanent structures. Koerner and Narejo (1995) used a modified version of the California Bearing Ratio (CBR) penetration test to investigate the susceptibility of GCLs to bearing capacity failure under concentrated loads.

---

[1]Assistant professor, School of Civil Engineering, Purdue University, West Lafayette, IN 47907.

[2]Graduate Research Assistant, School of Civil Engineering, Purdue University, West Lafayette, IN 47907.

Replicate specimens of three GCL products were hydrated for 24 hours in a 150 mm diameter CBR mold under a confining stress of 0.68 kPa. After hydration, the specimens were covered with a well graded sand of varying thickness H. The cover soil was then penetrated at a constant rate of 0.25 mm/min using a piston having diameter B = 50 mm. The force-displacement curves and visual appearance of the specimens were compared for H/B ratios of 0, 0.3, 0.5, and 1.0. On the basis of these tests, Koerner and Narejo recommended that an H/B ratio of 1.0 or greater is needed to protect a hydrated GCL from bentonite squeezing under concentrated loads. De Battista and Fox (1996) replicated Koerner and Narejo's tests for a needle-punched GCL. It was concluded from these tests that GCL bearing capacity may be dependent on the type of cover soil. Not all cover soils will exhibit a well defined peak strength as H/B increases. In addition, the experimental data suggested that a minimum H/B ratio of 1.5 may be needed to adequately protect an underlying GCL.

The objective of this paper is to present additional information regarding the CBR bearing capacity test for hydrated GCLs. A series of tests were performed to study the influence of product type, surcharge pressure, displacement rate, mold diameter, and termination criterion on the lateral squeezing of the hydrated bentonite. Based on this study, recommendations are provided regarding the procedure for performing CBR bearing capacity tests on hydrated GCLs.

EXPERIMENTAL PROGRAM

The procedure for the CBR bearing capacity tests consisted of three stages: 1) hydration of the GCL, 2) penetration of the cover soil using a piston, and 3) measurement of the GCL thickness profile. The geometry for the hydration stage is shown in Figure 1a. A circular GCL specimen was cut to fit the bottom of a confining mold having diameter D. A non-woven geotextile (previously wetted) and porous load plate were placed on top of the GCL, and a surcharge weight was placed on the load plate. To hydrate the GCL, tap water was ponded inside the mold and the mold itself was placed in a water bath. The perforated base plate of the mold also permitted the GCL to hydrate from the bottom. After hydration, the surcharge, load plate, geotextile, and water were removed from the mold, and a cover soil of thickness H was placed over the GCL. The cover soil was a poorly graded sand containing no gravel or fines, and had a mean grain size D_{50} = 0.6 mm. The geometry for the penetration stage of each test is shown in Figure 1b. The cover soil was penetrated using a CBR piston at a constant rate of displacement, r, and the piston force was recorded as a function of displacement. Depending on the objective for the test, penetration was terminated at a specified piston force or piston displacement. The cover soil was then carefully removed from the mold, and the thickness, t, of the specimen was measured as a function of position, x, along a diameter. To perform this operation, a ruler was laid across the top of the mold, and the distance from the top of the mold to the GCL surface was measured using a caliper (Figure 1c). Knowing the depth to the base plate of the mold, the thickness profile of the GCL could be calculated.

Table 1 provides a summary of the testing variables for the experimental program. Three types of commercially available GCLs were used in this study. Most of the bearing capacity tests were performed on GCL-1 (Claymax 200R). This product consists of adhesive-bonded bentonite between two woven geotextiles. Some comparative tests were also performed on two needle-punched products, GCL-2 (Bentomat ST) and GCL-3 (Bentofix). For these GCLs, the bentonite is held between non-woven and woven geotextiles that are needle-punched together. The needle-punched products were tested with the woven side facing upward. GCL specimens were hydrated at two values of overburden pressure, p = 0.69 and 6.9 kPa. The hydration time and piston diameter were fixed at 24 hours and 50 mm, respectively. Comparative tests were performed using three mold diameters and five piston displacement rates. The thickness of the poorly graded

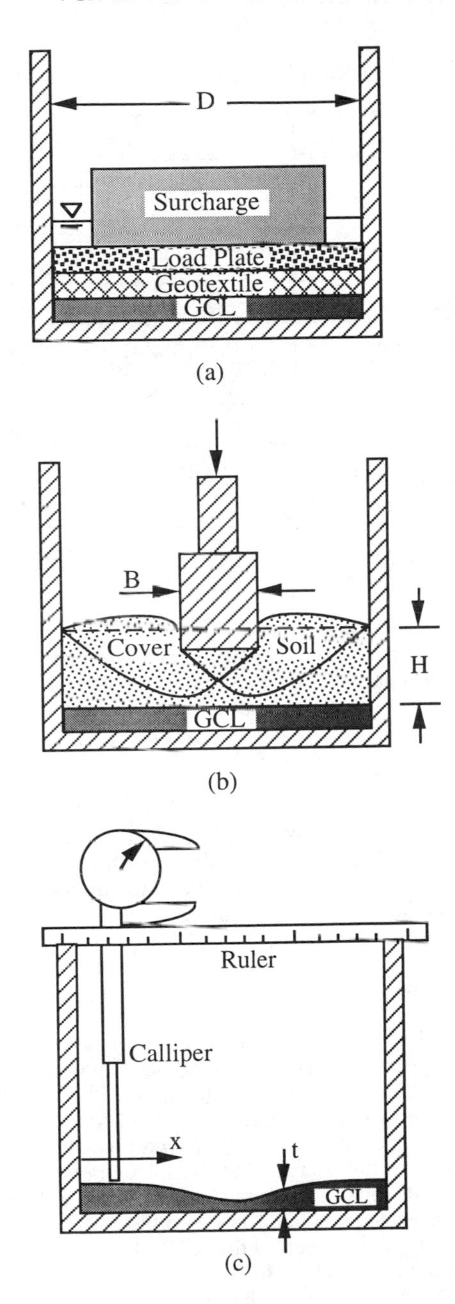

Fig. 1. CBR bearing capacity tests on GCLs: a) hydration stage, b) penetration stage, and c) thickness profile measurement.

Table 1. Summary of GCL bearing capacity experimental program.

Test Variable	Values for Experimental Program
GCL product type	Adhesive-bonded, Needle-punched
Hydration pressure, p	0.69 kPa, 6.9 kPa
Hydration time	24 hours
Mold diameter, D	152 mm, 235 mm, 305 mm
Displacement rate, r	0.12, 0.25, 0.5, 1.0, and 2.0 mm/min
Piston diameter, B	50 mm
Cover soil	Poorly graded sand
Cover soil thickness, H	50 mm
Cover soil thickness ratio, H/B	1.0
Termination criterion	Specified force, Specified displacement

sand cover soil was 50 mm, thus giving H/B = 1.0 for all tests. Both specified force and specified displacement termination criteria were investigated as part of the experimental program. Most tests were terminated at a specified piston displacement of 20 mm.

EXPERIMENTAL RESULTS

Individual series of tests were performed to investigate the effect of GCL product type, hydration pressure, displacement rate, mold size, and termination criterion on the lateral squeezing of hydrated bentonite under concentrated loads. Results from these tests are presented in the following sections.

Effect of GCL Product Type

A study of the bearing capacity of three GCL products was initially conducted in the experimental program. Specimens of each product were hydrated under an overburden pressure p = 6.9 kPa and tested in a 152 mm diameter mold. The force-displacement curves for these tests are shown in Figure 2a. For each product, a peak load occurs at 5 to 6 mm of displacement. After each peak, the load decreases slightly and then continues to increase until the tests were terminated at 20 mm displacement.

Although the force-displacement curves in Figure 2a are generally in close agreement, some differences can be seen. Based on the results of Koerner and Narejo (1995), the authors had originally expected that GCL-1 would show a less stiff response than the two needle-punched products. This is not observed for the initial part of the curves in Figure 2a. The curve for GCL-3 has a smaller initial peak, but then increases to nearly match GCL-2 at the end of the test. Conversely, the GCL-1 and GCL-2 curves are initially nearly identical, and then diverge as the GCL-1 curve falls below both needle-

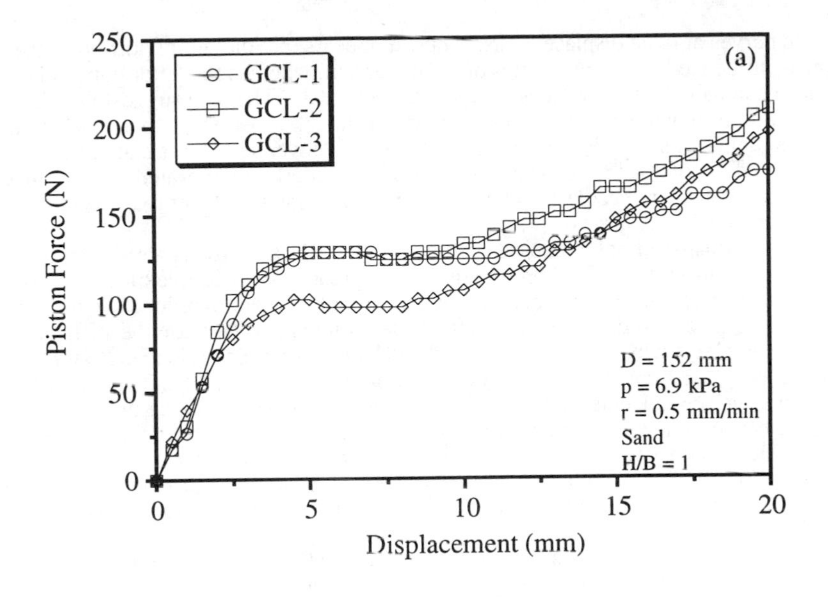

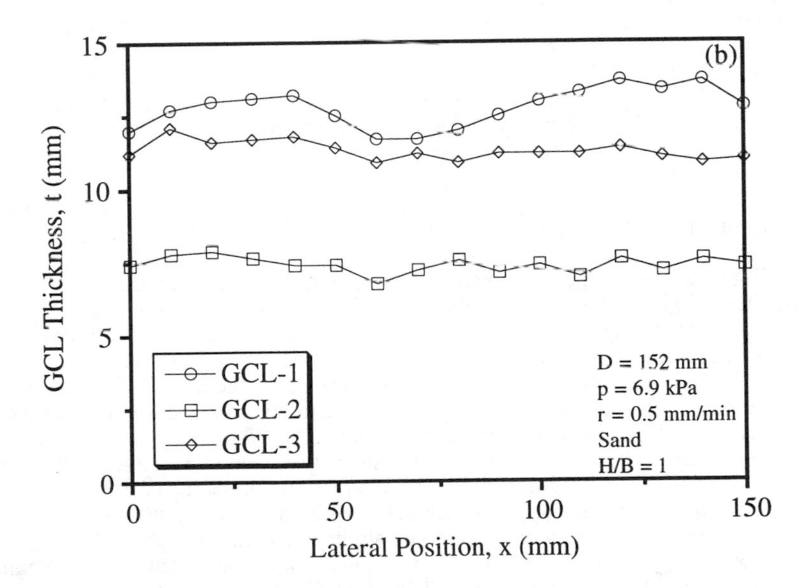

Fig. 2. Bearing capacity of three GCLs: a) force-displacement curves, and b) thickness profiles.

punched curves at large displacements. Thus, it appears that the initial portion of the curves is influenced by the properties of the cover soil, whereas the latter portion is primarily controlled by the stiffness of the hydrated GCL. The data suggest that the cover soil for GCL-3 may have been less dense than for the other two tests, thus producing a smaller initial peak stress. At the end of each test, the cover soil is in a state of failure and the measured piston load reflects the stiffness of the underlying hydrated GCL. Since both GCL-2 and GCL-3 are needle-punched, these products show a stiffer response than the unreinforced GCL-1 at large displacements.

The measured final thickness profiles for the GCL specimens are shown in Figure 2b. After hydration, the GCL-1 specimen had the greatest swelled thickness because it is unreinforced. The reinforcing fibers for GCL-2 and GCL-3 act to reduce swelling under the surcharge load, and thus also reduce the water content and increase the stiffness of the clay. This explains why the reinforced products had a larger piston load at 20 mm displacement. After completion of the penetration tests, GCL-2 and GCL-3 showed no discernible surface indentation, which suggests that bentonite squeezing did not occur under the concentrated piston load. GCL-1 showed a small decrease in thickness under the piston, which is consistent with it having a higher water content after hydration. It is interesting to note that, even though a small amount of bentonite squeezing occurred, GCL-1 still had the largest thickness of all three products after the penetration test was completed.

Based on the test results shown in Figure 2 and the data of Koerner and Narejo (1995), it was concluded that adhesive-bonded GCLs have a somewhat greater potential for bentonite squeezing than reinforced products. The reinforcement improves the bearing capacity of a GCL from two standpoints: 1) it decreases the hydrated thickness, which reduces the water content and increases the stiffness of the bentonite, and 2) it impedes the lateral movement of the bentonite. For a given set of testing conditions, it is likely that a reinforced GCL will perform as well or better than an adhesive-bonded product under a concentrated load. Thus, to be conservative, the remaining bearing capacity tests in the experimental program were performed using GCL-1.

Effect of Surcharge Pressure During Hydration

Figure 3a shows the force-displacement curves for two bearing capacity tests on GCL-1 that were identical except for the applied vertical pressure during the hydration stage. The corresponding thickness profiles are shown in Figure 3b. Both plots illustrate that the stiffness and bearing capacity of the hydrated bentonite increase with hydration pressure. In Figure 3a, the lower hydration pressure produced a soft GCL specimen which yielded a smaller piston force throughout the test. Clearly, in this case, the stiffness of the GCL had an important influence on both the initial and final segments of the force-displacement curve. Figure 3b shows that the same specimen had a larger initial hydrated thickness and experienced significant bentonite squeezing under the concentrated piston load. On the other hand, the specimen hydrated under 6.9 kPa shows a stiffer force-displacement response in Figure 3a, and only slight squeezing in Figure 3b.

Considering the test results shown in Figure 3, it is concluded that the magnitude of the hydration pressure is an important variable for GCL bearing capacity testing. A hydration pressure of 6.9 kPa was chosen for the remaining tests in the experimental program because this value is considered to be more representative of typical overburden stress levels for field installations.

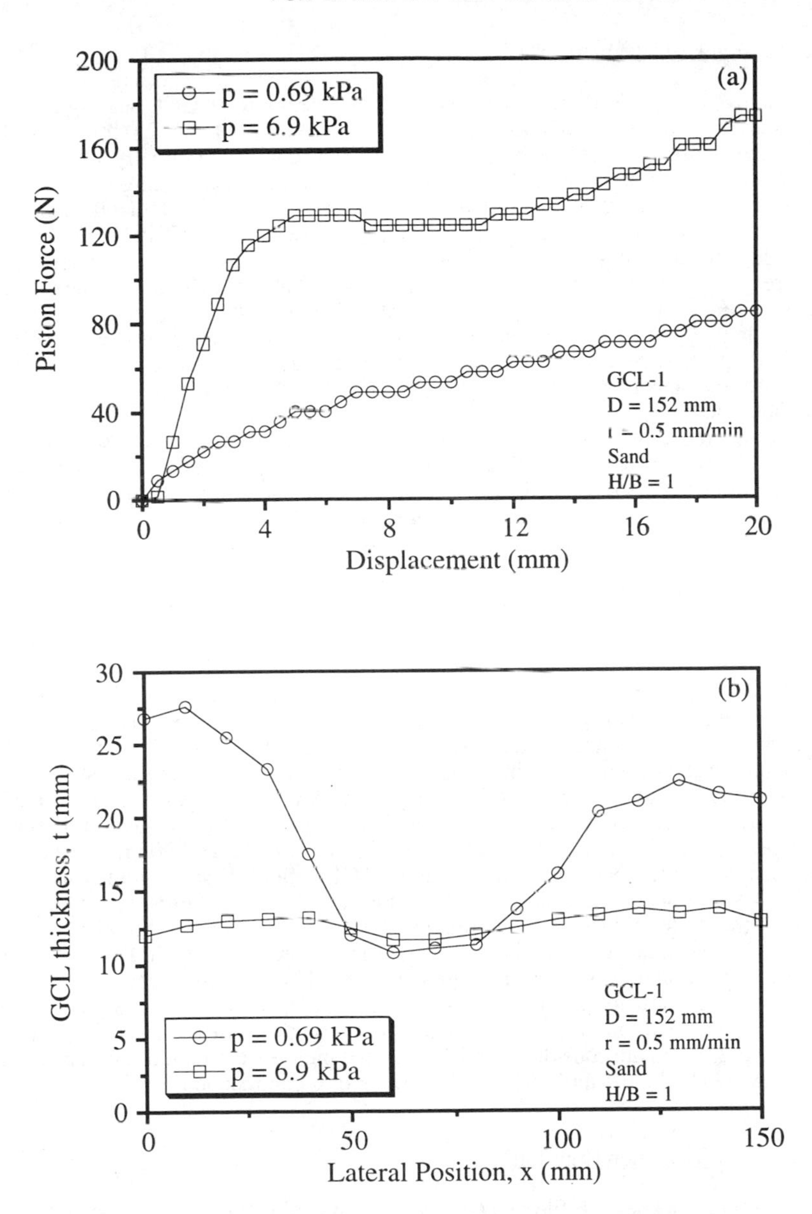

Fig. 3. Effect of hydration pressure on the bearing capacity of GCL-1: a) force-displacement curves, and b) thickness profiles.

Effect of Piston Displacement Rate

To investigate the influence of piston displacement rate on GCL bearing capacity, five tests were performed on GCL-1 for r = 0.12, 0.25, 0.5, 1.0, and 2.0 mm/min. Figure 4a shows the measured force-displacement curves. The curves are initially almost identical, and then, after 10 mm displacement, begin to separate according to piston displacement rate. In general, the measured piston load decreases with displacement rate. The effect appears to be more important for the slower rates (r = 0.12 and 0.25 mm/min), whereas the three tests at higher rates show nearly the same load at 20 mm. These test results are consistent with the well-established concept that the stiffness and shear strength of cohesive soils increases with applied strain rate (Taylor 1948, Bjerrum 1972). The corresponding GCL thickness profiles for the five specimens are shown in Figure 4b. None of the specimens show significant bentonite squeezing. Furthermore, increasing the displacement rate appears to have no significant effect on the surface indentation of the GCLs under the piston.

For the range of displacement rates shown in Figure 4, it is concluded that the rate of piston displacement has little influence on the bearing capacity of hydrated GCLs. Thus, the time required to perform a CBR bearing capacity test may be significantly reduced by increasing the displacement rate. The remaining tests in the experimental program were performed using r = 0.5 mm/min, which seemed to be a suitable compromise between fast and slow rates.

Effect of Mold Diameter

During a CBR bearing capacity test, the confining mold enforces a zero lateral strain boundary condition on both the GCL specimen and the cover soil which is not representative of field loading conditions. Thus, the mold should have a sufficiently large diameter such that its walls do not influence the measured response for the test. To investigate the effect of mold diameter, bearing capacity tests were performed on specimens of GCL-1 using three molds: D = 152 mm, 235 mm, and 305 mm. Shown in Figure 5a, the post-peak force-displacement curves are similar for D = 235 and 305 mm. However, the response for the test performed in the standard CBR mold (D = 152 mm) was consistently stiffer due to the additional lateral confinement. The corresponding thickness profiles for these tests are shown in Figure 5b. A similar small indentation was measured for each GCL specimen. Considering the larger piston force, one would have expected the measured indentation to be significantly greater for the D = 152 mm specimen. Since this was not the case, the stiffer penetration response for this specimen was likely due to the additional confinement provided by the mold. Based on this series of tests, molds having a diameter $D \geq 235$ mm are considered satisfactory for the measurement of GCL bearing capacity. A standard CBR mold (D = 152 mm) is not recommended.

Effect of Test Termination Criterion

The final thickness profile of a GCL specimen is directly related to the choice of termination criterion for a bearing capacity test. Piston penetration can be terminated at a specified force, specified displacement, or after a peak is recorded in the force-displacement relationship. For H/B = 1, Koerner and Narejo (1995) observed a peak in the force-displacement curves for three GCL products. On this basis, they concluded that, for H/B = 1, a general shear failure occurs entirely within the cover soil and the underlying GCL remains undamaged. De Battista and Fox (1996) failed to observe peak loads for their tests, some of which were continued to 40 mm displacement. Thus, it would seem that the

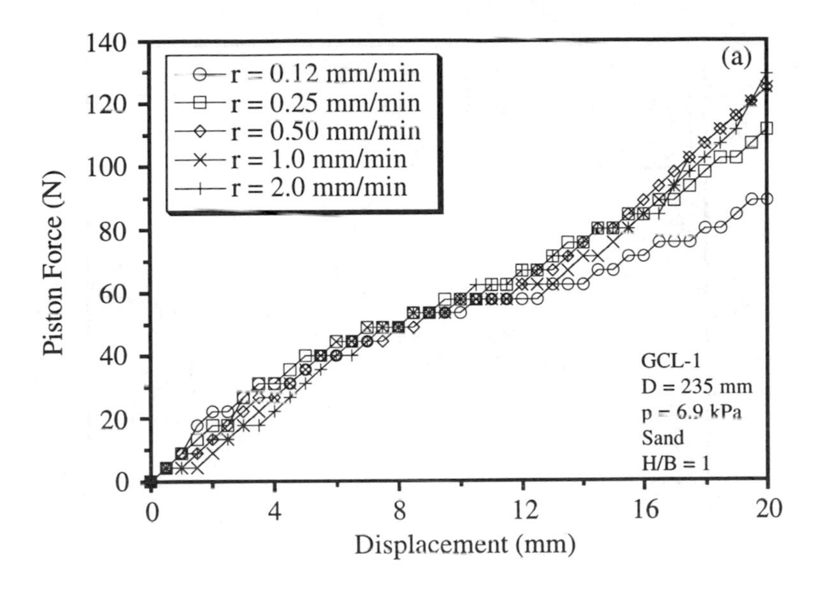

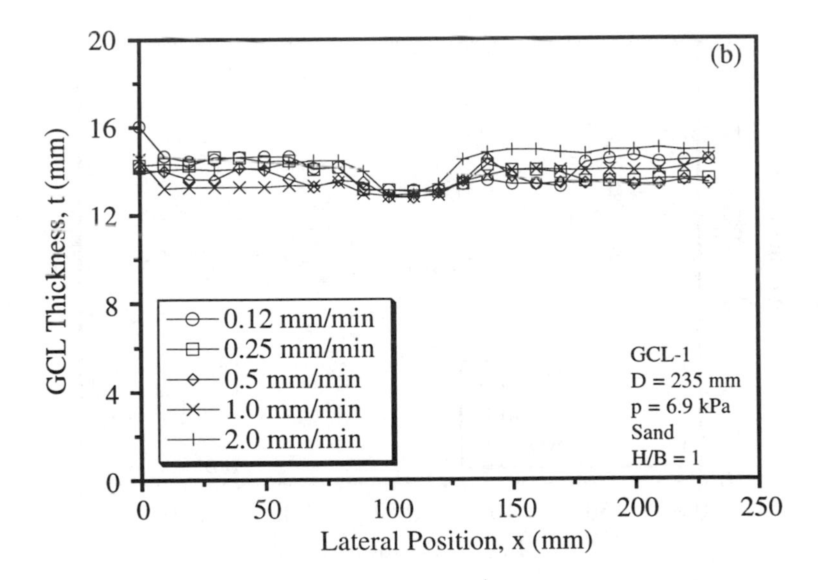

Fig. 4. Effect of piston displacement rate on the bearing capacity of GCL-1: a) force-displacement curves, and b) thickness profiles.

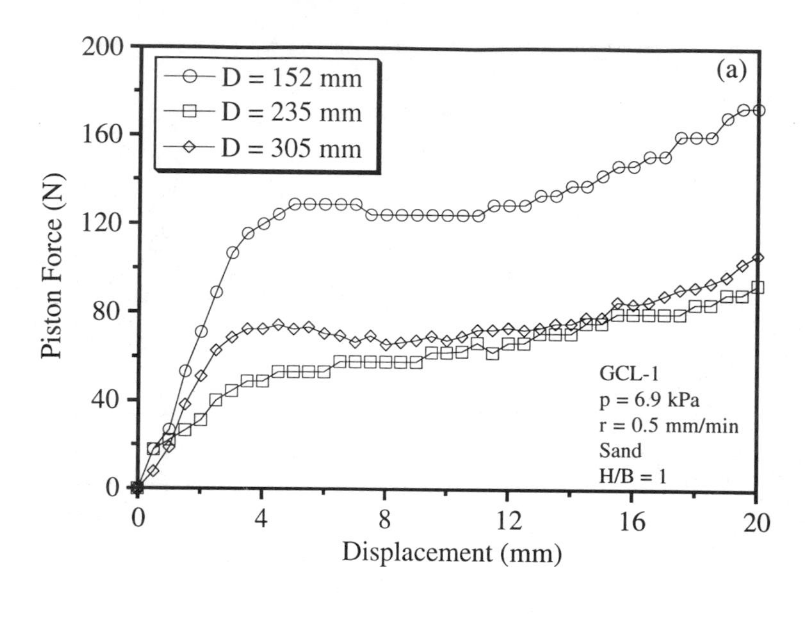

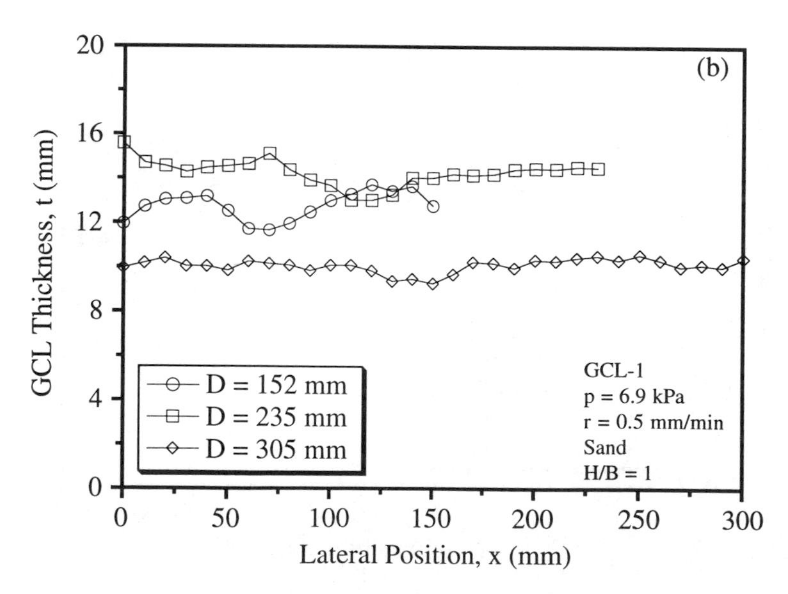

Fig. 5. Effect of mold diameter on the bearing capacity of GCL-1: a) force-displacement curves, and b) thickness profiles.

shape of the load-displacement curve may be dependent on the type of cover soil. Not all cover soils will exhibit a well defined peak strength as H/B increases. Consequently, a termination criterion defined based on a peak force may be unreliable in some cases.

Since peak strengths are not always observed in the measured force-displacement response, the choice of relevant termination criterion is unclear. Considering that field loads are generally constant stress (or force) rather than constant displacement, a series of six tests were conducted to investigate the applicability of a specified force termination criterion. Figure 6 shows the measured force-displacement curves for these tests. The first test, shown by the dark circles, was performed on the sandy cover soil with H/B = 3 and no underlying GCL. At 20 mm displacement, the measured piston force was 75 N. Using the same cover soil, bearing capacity tests were performed on replicate specimens of GCL-1 for H/B = 0.5, 1.0, 1.5, 2.0, and 3.0. Each test was terminated when the piston force reached 75 N. Using this termination criterion, Figure 6 shows that the final piston displacement decreases with H/B. Once uncovered, the measured thickness profile for each GCL showed no significant indentation due to the piston load. For these tests, the constant force termination criterion produced similar GCL indentation irrespective of H/B. It was therefore concluded that a specified displacement termination criterion, as used for Figures 2 through 5, is more appropriate for GCL bearing capacity testing.

For a specified displacement termination criterion, the amount of bentonite squeezing is a function of the piston displacement. Clearly, any GCL will be damaged if piston penetration continues indefinitely. Figure 7a shows the force-displacement curves for two replicate bearing capacity tests on GCL-1. The first test was terminated at 20 mm displacement, whereas the second was discontinued at 27 mm displacement. The force-displacement curves are nearly identical after the cover soil reached a failure condition (between 8 and 20 mm displacement). After 20 mm, the load for the second test continues to increase up to 27 mm displacement. Since the final load was larger for the second test, it

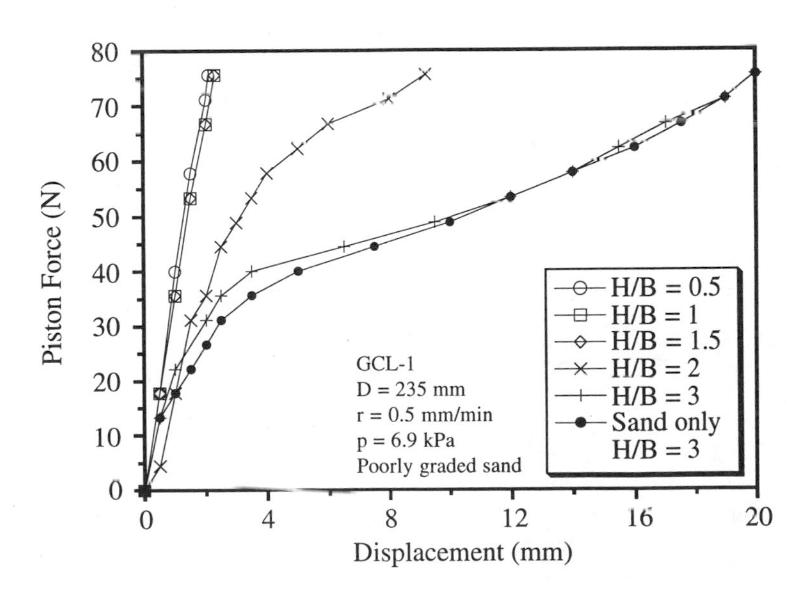

Fig. 6. Force-displacement curves using a constant force termination criterion.

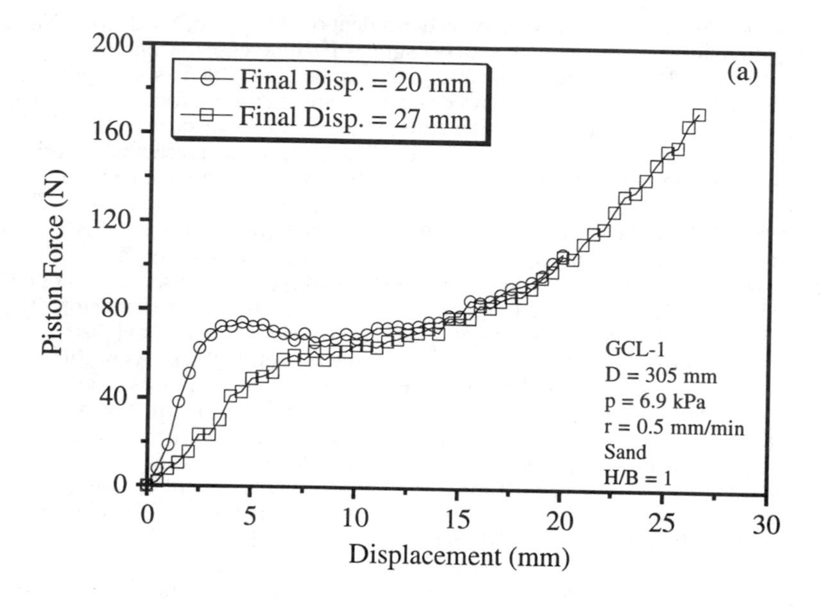

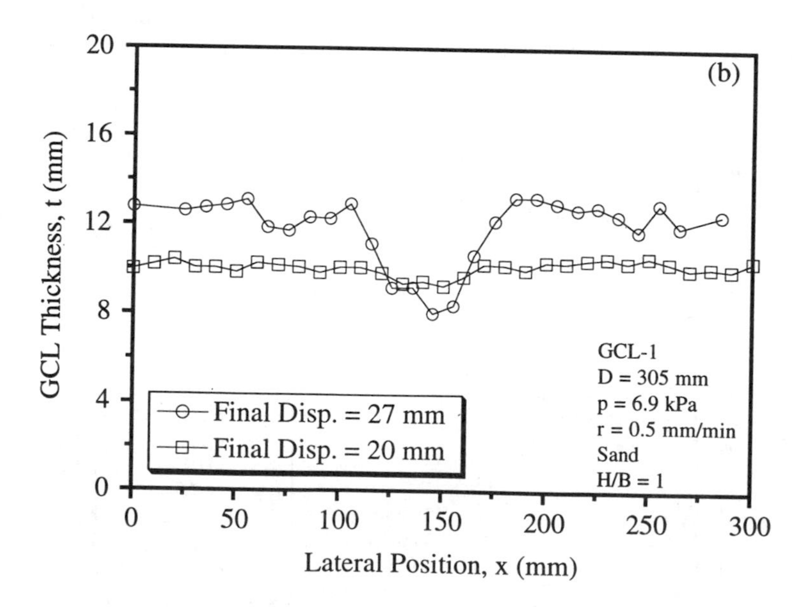

Fig. 7. Effect of final piston displacement on the bearing capacity of GCL-1: a) force-displacement curves, and b) thickness profiles.

would be expected that the GCL would show a larger indentation as a result. The corresponding thickness profile plots for these two tests, shown in Figure 7b, support this hypothesis. The GCL for 27 mm final displacement shows more bentonite squeezing. It is concluded that, for measured thickness profiles to be comparable, the final piston displacement must be consistent between GCL bearing capacity tests using a specified displacement termination criterion. However, the force-displacement curves may be compared even if the final displacement values are unequal (such as in Figure 5a).

SUMMARY AND CONCLUSIONS

The following conclusions are reached as a result of this study of the CBR bearing capacity test for hydrated geosynthetic clay liners:

1) Adhesive-bonded GCLs show a somewhat greater tendency for bentonite squeezing than reinforced GCLs under concentrated loads. Thus, for a given set of testing conditions, it is likely that a reinforced GCL will experience a smaller reduction in thickness than an adhesive-bonded product.

2) The magnitude of the surcharge pressure during hydration is an important variable for GCL bearing capacity testing. A hydration pressure should be chosen that is representative of the overburden stress expected in the field.

3) For the range of displacement rates investigated in this study, the rate of piston displacement has little influence on the bearing capacity of hydrated GCLs. A piston displacement rate of 0.5 mm/min was used for this investigation.

4) The CBR mold should have a sufficiently large diameter D such that its walls do not influence the measured response for a bearing capacity test. Molds having $D \geq 235$ mm were found to be satisfactory for the measurement of GCL bearing capacity. A standard CBR mold ($D = 152$ mm) is not recommended.

5) Termination criteria based on a specified force or a peak in the force-displacement relationship may be unreliable for some tests. A specified displacement termination criterion of 20 mm was found to be more appropriate for the GCL bearing capacity tests conducted in this investigation.

A number of additional variables, not discussed in this paper, may affect the results of CBR bearing capacity tests on GCLs. These include: piston diameter, piston shape, type of cover soil, hydration time, dynamic loading conditions (such as that simulating the passing of a vehicle), and creep effects for sustained loads. Some additional laboratory tests are currently being performed to characterize the effect of cover soil grain size on the bearing capacity of GCLs. However, full-scale tests are ultimately needed to assess the likelihood for local bentonite squeezing after the placement and hydration of GCLs in the field.

ACKNOWLEDGMENTS

Funding for this investigation was supplied by the U.S. National Science Foundation and the Colloid Environmental Technologies Company (CETCO). Geosynthetic clay liner samples were donated by CETCO and the National Seal Company. The assistance of Ahmed Karim with the bearing capacity tests for the 305 mm diameter specimens is greatly appreciated.

REFERENCES

Bjerrum, L. (1972). "Embankments on soft ground," *Proceedings*, Specialty Conference on Performance of Earth and Earth Supported Structures, Purdue University, American Society of Civil Engineers, 2:1-54.

De Battista, D. J., and Fox, P. J. (1996). Discussion to "Bearing capacity of hydrated geosynthetic clay liners," by Koerner, R. M., and Narejo, D. (1995), *Journal of Geotechnical Engineering*, American Society of Civil Engineers, in press.

Koerner, R. M., and Daniel, D. E. (1994). "Technical equivalency assessment of GCLs to CCLs," *Proceedings*, 7th GRI Conference, Industrial Fabrics Assoc. Int. Publ., St. Paul, Minn., 265-285.

Koerner, R. M., and Narejo, D. (1995). "Bearing capacity of hydrated geosynthetic clay liners," *Journal of Geotechnical Engineering*, American Society of Civil Engineers, 121:1:82-85.

Taylor, D. W. (1948). *Fundamentals of Soil Mechanics*, John Wiley & Sons, Inc., New York, 700 pp.

Author Index

Subject Index

Subject Index